The Disconnect

The Disconnect

A Personal Journey Through the Internet

Roisin Kiberd

First published in Great Britain in 2021 by Serpent's Tail,
an imprint of PROFILE BOOKS LTD
29 Cloth Fair
London
EC1A 7JQ
www.serpentstail.com

10 9 8 7 6 5 4 3 2 1

Typset in Freight Text by MacGuru Ltd
Designed by Barneby Ltd
Printed and bound in Great Britain by CPI Group (UK) Ltd, Croydon, CRO 4YY

A CIP record for this book can be obtained from the British Library

ISBN: 978 1 78816 577 8
eISBN: 978 1 78283 731 2

For my family, and for Rob

Dark Euphoria is what the twenty-teens feels like. Things are just falling apart, you can't believe the possibilities, it's like anything is possible, but you never realized you're going to have to dread it so much. It's like a leap into the unknown. You're falling toward earth at nine hundred kilometres an hour and then you realize there's no earth there.

Bruce Sterling, Reboot 11 speech, 2009[1]

Contents

Prologue: SWIM

I AM THE NEW FLESH. I live under technology, and technology is part of me.

I have stolen this term, 'the new flesh', from *Videodrome*, David Cronenberg's 1983 horror film, because it best describes how I feel. In it, people are altered by the media they consume. They mutate, shifting between euphoria and dread. The screen is so addictive, so hypnotic, that they return to it again and again, until it distorts their thoughts and threatens their humanity.

I too am addicted to the screen. Sometimes I think I have spent so much of my life online that I was raised by the internet. I've forgotten where the borders are, where technology ends, and where I begin. Am I a mutant? A cyborg? Or just an ordinary human?

A cyborg is a person whose physical abilities have been extended by technology. My extensions are not physical; my body looks much the same as that of someone alive in the pre-internet age. But I am an emotional cyborg. I outsource my opinions, my memories and my identity to the internet, and I have spent more time with my laptop than with any living being on earth.

This state is not unusual. I'm certain that the several billion other people in the world who use the internet experience it too.[1] Our use of technology is changing us, in ways we have yet to understand.

It was once claimed that the internet would liberate us. Techno-utopians claimed online life would allow us to transcend gender, age, class and race, and to construct our own identities. It hasn't turned out this way. Instead, we've been led into fixed identities, each person given a biography, a Timeline, and a filter bubble of their own.

Perhaps I am a techno-dystopian. Over the years, I've experienced a slow depersonalisation; cut off from reality, from sincerity and sensation, I've felt urged to compete in a scrolling world. I've been conflicted, at times wanting to be the ideal data subject, then distrusting technology, even as I turned its surveillance on myself.

Perhaps I'm not a cyborg, or a mutant, but a person split in two. As long as I've had a life in data I've also had a doppelgänger. Everyone has one: a shadow that exists in lists and systems, information stored on data farms, on servers hissing and blinking in the dark. The internet tracks us and pieces together a second self, and our every interaction with a service or platform adds to this profile, and is monetised by strangers. Even before you join Facebook, as one example, Facebook has already created a 'shadow profile' around you, a void waiting to be filled.

We live in data, yet we do not, because data is like dead cells shed from our bodies. The internet feeds us what it believes we want, based on what we wanted in the past. This means our doppelgängers are bland and predictable ciphers, the most narrow-minded version of ourselves.

Our doppelgängers grow stronger every time we use the internet. They will surely outlive us one day. They don't belong to us; artificial intelligence is the technology that will dominate our future, and it will be built on the data we create today, for tech companies.

Some claim we'll reach a point of Singularity in our lifetimes, an 'intelligence explosion' where machines will eclipse human capabilities for ever. It's described as the dawn of a sci-fi era, an apocalypse, or a release of near-infinite possibility. But how will technological, even spiritual revolution arise from an internet dominated by surveillance capitalism? Will the future be built on our futility and distraction, on what technology takes from us, as well as what it provides?

The Singularity will be boring, because the internet today is boring. The Singularity will not save us from a dystopia built by human hands.

I had to lose my mind before I was able to write this book. 2016, when the ideas for many of these essays began to take root, was also the year of my mental breakdown. I hadn't been very happy to begin with; since childhood, I'd gone through periods of depression, anxiety, eating disorders and self-doubt.

For years, the internet had been part of my life. I loved its strangeness, its creative possibilities, and this inspired me to write about it. Every niche, and every peculiarity of human nature was present online. I rarely felt like I belonged in the world, but I belonged on the internet.

I'd already been writing for websites and newspapers for several years, focusing on the intersection between the internet and real life. But now life didn't feel very real any more; most of it I spent alone with my laptop, watching a turbulent new culture emerge on the screen.

Throughout 2016, I watched as everything I found fascinating about online life turned dark. Subcultures were coalescing, self-radicalising, pooling their hate. Political beliefs gave way to black-and-white thinking, and anonymity permitted people to attack each other. My screen was a frenzy of tabs and feeds, clap backs, call outs, hot takes, doxxing, swatting, and people typing in ALL CAPS, silently shouting at each other in tweets. I'd scroll through it for hours; it hurt me, but I couldn't look away.

I knew I was already emotionally unstable, but now everything online was as extreme as my feelings. I saw all my fears confirmed as true; on the internet, we are being watched, not only by state agencies, and corporations, but also by each other. Your friends are all trying to make you jealous. Men do hate women, and women hate men, and yes, everyone *does* hate each other; maybe not in real life, but on the internet, which was beginning to feel like the same thing.

It took a while to realise that the internet had eaten my life. It took even longer to realise that I was experiencing a breakdown, because so much of the internet feels like a breakdown already.

Summer passed by, but my curtains shut out the sunshine. I sat behind the screen and watched Twitter, early in the morning and late into the night, as friends and strangers on my Timeline engaged in what is called 'the discourse'. I took on more work, writing articles each week along with advertorials, vast 3,000-word pieces featuring paid mentions of tech companies. I lived in a house on the Northside of Dublin with two friends, but barely left my room, and slowly became afraid of other people. I broke up with someone I was in love with, stopped eating and then stopped sleeping, and started going to the gym at night instead.

By the end of the summer I realised that I was numb. I was stuck behind a screen, and behind that screen, I was stuck inside a body that felt very little apart from exhaustion. There's an assumption on social media that we are consistent people, true to our Timelines, expressing the same beliefs and tendencies online as in real life. This could not be further from the truth, but at the time I believed it to be real; I thought everyone was utterly certain of themselves and what they stood for, and that my inability to be the same was a sign of some deep-rooted, incurable failing.

'SWIM' is an acronym for 'someone who isn't me'. It's used on forums dedicated to drug use, or mildly illegal acts like shoplifting, when the author wants to ask questions without risking self-incrimination. Usually SWIM will ask about dosage, or where to buy drugs. In the worst cases, SWIM will have been caught, and will ask for legal advice.

The use of SWIM is self-defeating, because the minute you see it you know that the author is up to no good. I was up to no good; I was reading about what I needed for an overdose. It didn't feel like suicide, because it didn't feel like I had a life any more. I would simply remove myself from existence, as easily as deleting an online account.

I waited until my housemates were away for the weekend, then I set my Twitter to private and my Facebook to invisible. I wrote out the passwords to my online accounts on a piece of paper, then swallowed about a month's supply of pills, and some painkillers just in case, and washed everything down with cheap coconut-flavoured rum. Then I lay down and felt my world turn into pillows, a soft, distorted nothingness which must be what it feels like to let go of Someone Who Isn't Me.

I lived, of course.

After the time off work, the group programme and cognitive behavioural therapy, I realised that I had lost perspective, but that writing offered a way to claim it back. I knew, also, that I couldn't simply blame technology for my problems, any more than I could blame other people, but that for me, and likely for others, the internet and mental health were closely intertwined.

I identified a change in myself, then I began to notice it all around me: distraction, loneliness, an ambient sense of existential crisis that is the internet's default state. We've adopted technologies that manipulate our emotions and limit our view of the world. Social media, especially, encourages us to type before thinking, or fact-checking, and to view life the way a machine categorises data – in binaries, leaving little room for complexity.

These essays were written during and about a process of recovery, but not withdrawal, from the internet. They are an attempt to make sense of what we've lost, and to consider the lonely dystopia in front of us. Donna Haraway writes that 'writing is pre-eminently the technology of cyborgs', and that 'cyborg politics is the struggle for language and the struggle against perfect communication, against the one code that translates all meaning perfectly'.[2] If I am a cyborg, I write in the spirit of this ambivalence, in defence of the imperfect and human. To write is to manipulate information, to claim it as one's own. This essay, and the book you hold in your hands, is the product of information mined from machines and human life.

Introduction: A History of the World Since 1989

I WAS BORN IN DUBLIN, the same month and year as the internet as we know it. In March 1989, an engineer named Tim Berners-Lee submitted a proposal to his employers at CERN for a new system of 'information management'.[1] It described a decentralised, open-source map of information, connected by hyperlinks.

'The WorldWideWeb (WWW) project aims to allow links to be made to any information anywhere,' Berners-Lee wrote in a Usenet post. 'We are very interested in spreading the web to other areas, and having gateway servers for other data. Collaborators welcome!' Berners-Lee's superiors responded by calling it 'vague, but exciting'.[2]

This technology advanced throughout my first years of life, and use of the internet gradually spread beyond academia and the military. In 1992, in a lab in Urbana, Illinois, the fictional computer HAL 9000 became operational. The first popular web browser, Mosaic, later called Netscape, appeared in 1993, the same year that id Software released *Doom*, inaugurating a gory new era of first-person gaming. A 30-year-old Jeff Bezos founded Amazon in 1994, naming his company after the longest river on earth, while in England a group of cyber hippies protested the law against outdoor raves by email-bombing the government and overwhelming their servers. It was the world's first act of online civil disobedience, and is remembered as the 'Intervasion of the UK'.

At the start of 1994, the websites accessible online numbered 623.[3] By the end of that year, the number had grown to over 10,000, with over 20 million internet users.[4]

I remember none of this. I was, after all, only five years old.

In August 1995, Bill Gates, then the CEO of Microsoft, danced awkwardly on a stage to 'Start Me Up' by the Rolling Stones.[5] He paid $3 million for the rights to the song. Gates wore high-waisted trousers and a polo shirt, his hair in a glossy bowl-cut, and as he danced he was surrounded by his doppelgängers: other early-middle-aged white men, also wearing polo shirts and dad trousers, some dancing more enthusiastically than others. In the outside world, beyond this campus in Redmond, Washington, people queued outside malls to purchase the product Gates was launching: Windows 95, the operating system that would make Microsoft a household name.

That same year, *The New York Times* and the *Washington Post* jointly published *Industrial Society and Its Future*, a 35,000-word treatise against technology, in a bid to stop the mail-bombing campaign instigated by its author. Linguistic analysis of the text helped police track him down, and in April 1996, Ted Kaczynski was arrested at his cabin in Montana. He was sentenced to life in prison, for taking arms against a wave of technological change even his extreme actions could not stop.

Momentum set in, and computers became an aspirational product for people who didn't already know how to use them. Netscape went public in 1995, launching its IPO without significant profits or revenue to speak of. eBay was founded, growing to 200,000 auctions per month in two years, in large part due to

Beanie Baby collectors. Hotmail launched the following year, and rapidly spawned users, signing off each message with its own viral marketing copy: 'Hotmail: Free, trusted and rich email service. Get it now.'

In 1996, a 23-year-old student named Larry Page created BackRub, a system of 'spiders' that crawl the web for links, arranging search results in an order he called PageRank. This marked the beginning of search engine optimisation (SEO), the value search engines assign to web pages and, increasingly, to the people they represent.

That same year, an article appeared in *Fast Company* by an American business writer called Tom Peters, titled 'The Brand Called You'. It outlined the future of cybernetic selfhood, a struggle for self-promotion where people market themselves like companies. Peters wrote:

> You're branded, branded, branded, branded. It's time for me – and you – to take a lesson from the big brands, a lesson that's true for anyone who's interested in what it takes to stand out and prosper in the new world of work.

You are your website, and the success of that website determines your worth. Peters blurred the lines between commerce and personhood, combining marketing advice with a near-mystical faith in cybernetic individualism. Employment rights got no mention here; Peters suggested working for free in return for self-promotion, and readily embraced cloud feudalism – in his vision of the future, we'll rely on internet platforms to keep us in steady, if temporary, work.

Personal branding, for all its hyperbole, is not about glory; it is about simply staying afloat. It aims to make of its reader the perfect data subject: the more you give of yourself to the internet, the more, apparently, you'll get back.

*

My parents didn't own a computer until 1996, the same year 'The Brand Called You' was published, when my father brought home an Apricot PC from work. The monitor was boxy and white, and the system unit was comically large compared to today's machines. I didn't use it much, but I knew how to play Minesweeper, and how to draw things with Microsoft Paint. Our babysitter, an older boy who lived next door, ran MS-DOS on it sometimes, and I remember the otherworldly look of the blue and the white, the stiff, typewriter-esque font, and the unsettling feeling that we were seeing the machine's entrails.

A few years later my parents upgraded to a giant, wheezing desktop made by HP. Soon after that we got dial-up. Readers alive in the 1990s will recall precisely the sound of the modem, the clunky melody of circuits, a mystic handshake between machines.[6]

Google was incorporated in 1998, its name a play on the number 'googol' – the digit 1 followed by 100 zeroes. That same year, the watch company Swatch announced an ambitious experiment in physics-based marketing: 'Internet Time', a concept that divided the day into 1,000 'beats' across time zones. It didn't take off. Netflix launched as a mail-order DVD rental business, Apple released the iMac, and a now-defunct electronics company, Diamond Multimedia, released a $200 device, box-shaped and roughly the size of a deck of cards, called the Rio PMP300, which became the first commercially successful MP3 player.

As the new millennium approached I began to explore the internet, which didn't feel limited then, even though it was. Users were staking out territory, creating homepages decorated with 1337 h4xor slang and animated GIFs. It felt exciting and vaguely illicit, priced by the minute and delivered in slow, guilty quantities. Someone elsewhere in the house was always waiting to make a phone call, and you were adding minutes to the bill, so every click needed to count.

Perhaps this is why my earliest online memories retain a deviant quality; pictures downloaded slowly, torturously, and websites that were deeply, sometimes inappropriately, personal. On Comic Chat I spoke with anonymous adults, and other children, from around the world in the guise of a cartoon beatnik, or an alien. One especially vivid memory is of the day of the porn virus. Our babysitter clicked on a bad link, or perhaps he was surfing dodgy websites, and the computer downloaded malware that manifested in video pop-ups. I remember watching the screen fill with gyrating actresses, ladies of the digital evening, and lipstick lesbians kissing in the back of a car. A week or two later, my parents hired a repairman to clear the virus away. That was the first time I thought of the internet as dangerous, a viral entity, waiting to infect you with one wrong click.

It didn't put me off: I pushed further into the internet alone, and at roughly the age of eleven I found pro-ana websites, which offered tips for 'perfecting' anorexia, a problem I didn't yet know I had. Those sites had a very 1990s look to them: black backgrounds with white text in Papyrus and Jokerman, butterfly motifs, sidebars full of bad poetry and starvation tips. They were lonely places, documents of suffering unspoken in the world outside the screen.

I don't remember Y2K, except for a juvenile thrill at the thought of a shiny new post-modernity, the earth overthrown by robots.

Of course, the humans behind technology were dangerous enough on their own. The first years of the new millennium saw the dotcom bubble burst, an ending less glamorous than anything imagined by prophets of technological doom. Pets.com came and went. eToys went bankrupt, leaving children without Christmas presents. An online currency called Beenz appeared, then disappeared, and was forgotten, while Pixelon, a company claiming to have created a revolutionary new TV-to-internet product, threw the legendary iBash '99, featuring performances by The Who, the Dixie Chicks, Faith Hill, Tony Bennett and Kiss. It cost over $16 million, more than 75 per cent of Pixelon's funding. Soon after, their CEO Michael Fenne, known for his volatile management style, was revealed to be David Kim Stanley, a conman and fugitive named on Virginia's most-wanted list.[7] His video company was secretly running Windows Media Player instead of its own, non-existent product.

At the end of the year 2000, when I was about to turn twelve, I received my first phone as a Christmas gift: the Nokia 8210. Most of my classmates had the 3210 instead, known for its changeable fascias – the one I remember boys at school owning had a picture of Eminem on it, crouched down, wearing a hoodie and a scowl – but the 8210 was lighter, smaller, and had appeared as an ad placement in the *Charlie's Angels* reboot earlier that year.

I quickly became attached to my phone. I decorated it with a pink hand-strap and Hello Kitty stickers. I collected polyphonic ringtones and odd, sentimental chain texts from friends I met at summer camp. Texting was itself a kind of pre-teen performance of independence; during long car journeys, on holidays and even during meals with my family, I would produce my phone – conspiratorial at first, under the table, but later shamelessly – and lose myself in composing the perfect text. This was my first taste of immersion in a tiny screen, a way to disappear in plain sight.

In 2001 the iPod launched, later joined by the iPod Mini. A cousin gave my brother and me his old one, filled with illegally downloaded tracks by Queen and Wu-Tang Clan. That same legendary cousin also gifted us his old PlayStation 2 and a copy of *Crash Bandicoot: The Wrath of Cortex*. Soon after that we got *Grand Theft Auto: Vice City*, and it taught us all about adult life.

That was the year HAL 9000 ran rampant, perniciously, fictionally, in space. The number of CDs burned worldwide became equal to the number sold in record stores. Google passed 100 million searches per year, or 1,000 queries per second, and began to place pay-per-click ads among their search results.[8] Apple launched iTunes, finally, and brought downloadable music into the mainstream.

A rash of social platforms appeared in the years that followed, including Flickr, OkCupid and MySpace. MySpace taught me a lot about hair products, and very little about musical taste. In school I pretended I was too cool for Bebo, but at home I signed up under a series of fake names to spy on my friends. For a while I wrote a LiveJournal about my feelings, then I signed up to Blogspot, where I posted pictures of my attempts to bake bread (not that I ate much of it; I was still anorexic, on and off, channelling anxiety about school exams into a diet that was almost as rigidly controlled as my study schedule). Neither of my blogs lasted very long; the posts were mostly apologies, made to non-existent readers, for not writing more often.

In January 2004, Mark Zuckerberg, a 19-year-old student and future Harvard dropout, registered thefacebook.com for $35 and launched the site from his college dorm room. By the end of that month, three-quarters of the student body checked in on it every day.[9]

In 2007 Facebook expanded to England through Oxford, Cambridge and the University of the West of England. One year after that, I arrived at Cambridge as an undergraduate. I resisted a while, then joined out of fear that I was missing party invitations and official college announcements. I remember setting up my Facebook page, and using the network for the first time. It was like creating a blog, but lazier; the user was asked to fill a prewritten template with information, rather than building one of their own.

In 2008, 145 million people worldwide had signed up to the social network,[10] Zuckerberg was the world's youngest billionaire, and Facebook opened an office in Dublin. The following year, at a rave somewhere in California, somebody who looked extremely like Zuckerberg was photographed dancing in front of the DJ booth, sweating from the face, eyes glazed and possibly high, or, at least, lost in rapture at the thought of taking over the world.[11]

When I graduated from college in 2010 I ported my college emails over to Gmail, and Google became my digital shepherd into adult life. Gmail was where I sent out CVs from, trying desperately to make myself seem grown-up and professional. Google Docs was where I worked, writing articles in a series of internships at print media companies, most of them already on the verge of bankruptcy. Gchat was also where my first serious relationship played out, in a series of flirtatious, then affectionate, then finally passive-aggressive sidebar chats, archived by Google for ever.

Lost in a normie shuffle, in June 2010, Steve Jobs danced on stage to a song by Jonathan Mann, a musician known for posting a new song every day to his widely followed YouTube channel.[12] After the song ended, Jobs announced the launch of the iPhone 4, the first iPhone to feature a front-facing camera, which made selfies easier, more popular and, ultimately, socially acceptable.

Elsewhere Netflix killed video shops, an 'app goldrush' was declared, and Bitcoin was invented by a pseudonymous genius. I remember attending a party around this time, the summer after graduation, where a guy offered me a joint from what looked like a selection box filled with different strains of weed. The Dread Pirate Roberts had recently launched his deep web marketplace, Silk Road; as I took a drag, my new acquaintance told me he'd signed up for a monthly subscription, having his drugs sent to a false address.

I spent a year in Dublin, writing fashion features and bad music reviews, breaking up and getting back together with my boyfriend, and steeping myself in blog culture, which at the time mostly involved the music site *Pitchfork*, the Cobrasnake, a widely mocked, much-imitated club photographer, and Tao Lin, the author, who doubled as a kind of career internet troll. I also remember reading the satirical blog Hipster Runoff, which may or may not have also been authored by Tao Lin.[13] Finally, my relationship ended, properly this time, and I decided to get out of Dublin.

It was in a market heavily influenced by personal branding that I began my working life after university, at an advertising agency in London. I was employed as one of an early wave of social media specialists, professional magpies employed to curate 'shareable content' and to 'drive engagement' between brands and their customers. I rented out my own personal brand – my voice, my taste, and my familiarity with internet subcultures – to companies that had little business being online in the first place. This was a time in which online 'customer interaction' very often went too far, relying heavily on hashtags, pandering to memes, and almost always coming off as disingenuous and smarmy.

During this time I lived in a dingy flat in Hackney, earned £22,000 per year, and believed that I had finally grown up. I was a Social Content Creative, working for an electronics brand as the moderator of a group of lighting engineers on LinkedIn, and, more regularly, as the custodian of social media accounts belonging to a popular brand of cheddar cheese. Consulting the data division, I would work out the best day and time to post on Facebook and Twitter, studying the habits of bigger, more successful brands and downloading endless white papers in order to work out how to phrase a 'killer call to action'. I would diligently write calendars of tweets one month in advance, and send them to the client for pre-approval.

In my cheese-related work, the words 'healthy', 'indulgent' and 'comfort food' were banned, and we were encouraged to use the terms 'on the go' and 'snacking'. The posts that did best tended to emphasise British heritage, and featured macros of melting cheese on toast. This particular cheese is not difficult to guess – a brand of supermarket cheddar, unchallenging and mellow, family-friendly and popular today, I'd imagine, with pro-Brexit voters.

My time as a cheese on the internet was eye-opening, a primer in the crude, instantaneous logic of social media. If you ask your followers to do something – 'Like', 'share' and 'subscribe' – there's a very good chance they'll do it. Followers respond well to competitions, giveaways and jokes, and are grateful if you take the time to reply to them. Crowdsourcing was a trend in digital marketing at the time, a response to a perceived democratisation of the media. People were optimistic about the role technology might play in building a better world, bringing transparency to politics and giving a voice to the marginalised. This was the early 2010s, a time when a blind faith in social media fuelled by Occupy, the Arab Spring and, less eminently, the campaign to find Ugandan war criminal and viral sensation Joseph Kony led many to believe that the combined powers of social media users could accomplish pretty much anything ('We did it, Reddit!').

This clearly isn't the case, because on the internet people are bored; they go there for distraction from work in the daytime, and at night from life itself. On Twitter, my cheese was followed almost exclusively by teenage girls, who befriended me only because the cheese was followed, likely as a joke, by Harry Styles, the milquetoast Mick Jagger impersonator then known for fronting the pop group One Direction. The girls often sent my cheese account private messages, begging me, or, rather, begging the cheese to contact Styles on their behalf and ask him to follow them. In a bid to get closer to this goal, they'd flatter me (or, rather, flatter the cheese), liking my bad puns and promotional images. Brands offered them some tiny amount of online cachet, one extra follower, and the training wheels for real social media interaction. On Facebook, meanwhile, I targeted the mothers of those same girls – my follower base was women in their thirties and forties, exhausted and charged with assembling school lunches every day, who took to branded social media pages to complain or to ask about 'BOGOF' offers at their local Tesco (it was over a month before I realised that this stood for 'buy one get one free').

In 2011 I acquired my first smartphone, a very basic Alcatel model paid for by the company I worked for. It was given to me in the understanding that I was now a '24-hour creative', someone who would reliably bring work home, reply to emails and check in on the cheese accounts at night and at weekends. The smartphone was a Trojan horse; it meant that I carried my job in my pocket.

My main memory of that job is of staying late and coming in early, and of vomiting up my lunches in the bathroom. I was a practising bulimic back then, turning my self-loathing into self-harm instead of properly addressing it. At the ad agency, the longer I watched money circulate at a distance – budgets spent on cheese tweets, instead of something, anything, more meaningful – the more my work felt like a dark joke.

Later I convinced the company I worked for to send me to South by Southwest, a tech conference held in Texas, where I delivered a panel on social media and the fashion industry. I witnessed some interesting talks and a host of sub-TED Talk blather, including *The Four-Hour Work Week* author Tim Ferriss discussing biohacking, which seemed to involve applying magnets and lasers to his head every morning to 'charge up' the brain cells. On the last day of the festival I got drunk at a Google-sponsored bar – Google had in fact rented an entire neighbourhood which they were calling the 'Google Village', with a corporate keg party in every house. I got talking to one of the barmen, and ran off with him to a dive bar down the street, resurfacing the next morning in a rickshaw.

That was the beginning of my corporate defection. I had almost begun to aspire to social media 'thought leadership', but that felt increasingly meaningless. My job as a cheese had lost its novelty; the joke wasn't even funny any more.

Back at work in London I found myself bored, and distinctly aware that I was taking advantage of my audience rather than connecting with them. There was something disturbing about the way I reached them; depending on how much money the client wanted to throw at their campaign, they could essentially pay for interactions or exposure to a large segment of cheese-eating Facebook users. Each meeting with the data team left me feeling oddly haunted; we were able to target absurdly specific groups of people by location, age, friendships, relationships and spending habits, invading their feed with cheese-related news as though they had elected to be the brand's 'friend', even though they hadn't. It felt like trafficking in humanity: pie charts and figures, and users themselves used as advertising. That year, Facebook announced that they would be allowing advertisers to use people's Likes and pictures to advertise to their friends, while Twitter, popular but lagging behind Facebook with 100 million active users,[14] announced targeted, location-based advertising for desktop and mobile.

The social media platforms called the shots, and would soon eclipse advertising and traditional media entirely in terms of power and influence. No one could question them, because everyone relied on them for clicks. Clients would give us money, and in turn we'd give money to Facebook and Twitter to access the data they gathered. It felt like a glimpse of the God View that social media platforms maintain over their users; we too could reach them, but only if we were prepared to pay the platforms for access.

After roughly a year I quit my job and moved back to Dublin. I'd started to have nightmares about being bullied by 12-year-olds on Twitter, and wondered if they guessed that I wasn't really a cheese, or that behind every account is a junior exec, or an intern, who struggles to make sense of a generation not much younger than they are. I also saw something exploitative, and potentially volatile, in the online 'crowdsourcing' encouraged by advertisers; through all the tweets, the 'community engagement' and 'customer empowerment', we charmed the public into doing our work for us, harvesting their data while pretending to be their friend.

Witnessing this new media landscape first-hand convinced me that online advertising, and the internet, and capitalism itself were utterly broken. I left my job with a sense of shame for having been part of this world, but I was also eager to unravel its workings.

That same year, on 22 September 2011, at Facebook's F8 developer conference in San Francisco, Mark Zuckerberg appeared on stage in front of a projection of his own profile page and announced the introduction of the Facebook Timeline. It would be, he said, 'the story of your life ... all your stories, all your apps, and a new way to express who you are'.[15]

On the Timeline, the events of each user's life would fill out a linear, scrolling path, detailing interactions on Facebook in chronological order, with the option of adding backdated posts and 'life events'. It drew inspiration from the quantified self, a tech-based movement which seeks 'self-knowledge through numbers', but it was also indebted to 'The Brand Called You', encouraging people to create personal pages on Facebook the way Tom Peters had once urged them to build websites.

Facebook framed the Timeline as a way to make their platform more personal, a place to bring together the disparate elements of the online self. It was also, more subtly, a way for them to lay claim to that same life's experience, assuming that everything could be recorded and processed as data with 'the whole story of your life on a single page'.

Like 'The Wheel', the photo projector carousel pitched by Don Draper in *Mad Men*, the Timeline invites the viewer to make of their messy past a more consistent picture, presenting a succession of memories steeped in the gentle luminescence of the screen. It's about imposing order, and narrative, on personal history by giving it to an online platform. The difference between the Facebook Timeline and Draper's carousel is that one is circular, while the other is linear. Both, however, can go on potentially as long as the human lifespan, bending the intricacies of experience to their shape.

Now Julian Assange is dancing.[16] A pale figure in darkness, lanky, blonde hair luminescent, arms flailing like branches on a broken tree, he makes his way across a dance floor in Reykjavik to merciless techno, a pounding beat, and lyrics about losing control. It's the early 2010s, and soon everything will change again. He'll be placed under arrest, he'll move from country house to Ecuadorian embassy, hero to villain, free man to prisoner, and the culture of the internet will change, dramatically, alongside him.

But for now Julian Assange is frozen in time, dancing at the door of oblivion. 'Would you believe that I still love you?' the lyrics demand. 'Would you believe?'

This is where my own Timeline starts to bend and break apart. After I quit my job, in 2012, I came home to Dublin and spent a year doing an MPhil at Trinity College. I wanted to go into academia, but near the end of that year I began to pick up writing work online. One piece I wrote did well on Twitter, and I gained around 1,000 followers in a single day. I experienced a small rush of hope and possibility, enough to believe that I could build a career on writing about the evolving – but, to my mind, essential – subject of life with technology.

In the years that followed I pursued that goal, navigating an online culture that was, though I didn't know it at the time, edging slowly towards a precipice. With Gamergate, the alt-right and other online mobs throughout the 2010s, that same sense of empowerment the advertisers exploited gave way to one of entitlement. Social media became the site of a culture war, one where thinly veiled feelings of futility translated into discrimination and threats of violence. It was no coincidence, to me, that these movements shared a distrust of mainstream media, while failing to recognise that platforms like Facebook and Twitter *are* the mainstream media today.

Social media users are dependent on the very thing that stifles them: the filter bubble, which skews and disguises the truth. The bubble closed in, gradually, throughout the 2010s, a decade which, ironically, began with calls for a new era of transparency in governments and in media. Wikileaks was part of this shift; everything could be documented online now, and subsequently every lie could be exposed.

Today it is obvious to me that this drive for transparency was hijacked, and led us into the arms of surveillance capitalism. If the 'democratised' online media is Facebook, then as a society we have settled for less. It's dangerous to assume neutrality of technology. It's dangerous not to scrutinise the people who build it. It's dangerous, also, to believe that simply by having the tools to change the world – social media platforms – we will automatically make the world a better place.

*

After three or so years of journalism I was given a column on internet subcultures, where I'd write about a new one every week. I felt honoured, but I also failed to anticipate that 2016 would be one of the most tumultuous years in the history of social media, a period of animosity, tribalism and unrestrained rage in which the darkest elements of online life would creep off the screen and into reality. Gradually, writing about online subcultures came to mean writing about culture itself. Still I continued, trying to make sense of a dark turn that had yet to manifest itself in full.

That year I interviewed Twitter trolls, YouTube trolls and Reddit trolls. I spoke to the nascent alt-right, interviewing fascist bronies and a far-right kid on Tumblr who posted pictures of himself wearing lederhosen. Their bizarre views spoke less of a coherent political ideology than of an outright loathing for humanity.

I worked hard, and didn't leave the house very often. I spent more time in front of a screen than ever before, and I spent more time reading people's tweets than having conversations face to face. A distance formed between me and the rest of the world, and I began to view reality in terms of the internet, where before it had been the reverse. Often I wondered if my interview subjects would end up becoming school shooters – young men on killing sprees appeared so often in the headlines that year that I lost count.

Exposure to these things affected me deeply, but I couldn't look away. I wrote about young people who used the Q&A platform Curious Cat to send themselves hate mail, because haters are more of a status symbol than friends. I also wrote about people who believed they were being 'gangstalked': followed, surveilled and tormented with mind-controlling technology by an unknown foe. Afterwards, they bombarded my Twitter with messages, some grateful, but more often aggressive, conspiratorial, or accusing me of being part of a campaign to harass them.

Then I wrote a piece about 2b2t, widely regarded as 'the worst place in Minecraft'. Unsure what to expect, I logged on and found myself adrift in a digital hellscape; apparently infinite terrain strewn with collapsing towers and pixelated swastikas, and populated with murderous players, who I assumed were mostly teenage boys.

A kind of emotional entropy set in; I withdrew from the real world, which seemed almost painful in its brightness, its demands to appear healthy, and balanced, and to sleep at night and work during the day. Instead I lived online, where everything I wrote, read and experienced fed into a paranoid mindset – the internet's, and my own.

The internet instils in us a conspiratorial worldview, the kind of maniacal self-involvement that comes from seeing friends on a list described as 'followers', or as a buffet of conversations waiting to be picked up or put down. It's there in the sense of competition that comes from having your every pronouncement rated with Likes and shares and retweets, the simultaneous self-importance and hopelessness that comes from thinking of yourself as a brand. It's the madness of treating a brand of cheese as a 'friend', the internet's illusory intimacy – the other side of surveillance – which leads you to believe your every thought must be made public.

On the internet you never see yourself on someone else's friend list. You experience life from a position of cybernetic solipsism, a filter bubble which reinforces all that you already know, however misguided, or biased, or unhinged it might actually be. Today I'm still shocked that this led me to attempt suicide – an act augured by months of distorted thinking.

What purpose does a Timeline serve? Is life so straightforward, so easy to narrate?

After the overdose, I ended up in a six-week outpatient programme in suburban south Dublin, a re-education plan for those who had forgotten how to be human. It was my first break from social media and email since I started using them.

Each afternoon in group sessions, we sat in a circle in vast, reclining La-Z-Boy chairs and described our particular rifts with humanity. Often, I noticed, people blamed the internet for their problems. Younger patients mentioned the pressures of social media, and older people said that the young were too lost in their phones, and had become vapid and cynical, that the old world of manners, consideration and kindness had fallen away for ever.

Perhaps the internet had sent me there, or perhaps I had been unavoidably unwell in the first place. The doctors diagnosed me with borderline personality disorder, a condition characterised by emotional volatility, black-and-white thinking and an unstable, frequently absent sense of self. Perhaps, then, it was a collision of a broken person with a broken and highly addictive system.

It was only at the end of 2016 that I noticed a change made to social media, one that helped me understand why this had been a year like no other. The Timeline was warped; it was no longer chronological. That year, Twitter, Facebook and Instagram replaced the scrolling newsfeeds users saw when they logged in with algorithmic timelines, tailored to each individual. Now, instead of seeing the latest stories each time we logged in, we were given a version of events curated by the platform.

That year saw a significant change, not only to online culture but to how its stories are told. Technology rewrote time, and told us only what we wanted to hear, and in the process, I believe, it helped to destabilise the truth. It seems little coincidence to me that this was also the year of 'fake news', a period in which troll farms and data mining interfered with democracy. It was also a year in which voices got louder, opinions and actions became more extreme, and online society, overall, began to compete for the attention of these same algorithms.

Something strange happens when we're offered an outlet to express ourselves, but only on the condition of knowing that we will be one among millions, even billions of voices competing to be heard. It brings out the extreme; it sends the mind into a state of panic. Too often we're encouraged to seek self-worth online, even if we're also warned, almost as frequently, that the pursuit is meaningless. I think this forlorn promise was what attracted the people I interviewed – the conspiracy theorists, the trolls and extremists and lonely teenagers – to their chosen medium, the internet. It offered them an identity.

What is humanity to a machine? What do machines do to our humanity? Each day we're faced with a digital sublime, but our dealings with it are limited to search boxes.

There's a meme I've noticed on Twitter and Tumblr. It's the painting *The Wanderer Above the Sea of Fog*, reconfigured for the internet age. Caspar David Friedrich's vision of a young man standing on the edge of a precipice is altered, so that now he's in front of a Google search bar, or a scrolling Tumblr timeline. He's lost in contemplation, and, like every internet user, forever alone.

The Wanderer GIF speaks to the existence of a digital sublime. On some level we know that when we interact with technology we confront our own powerlessness; we address our looming obsolescence, and, before that, our dependence on machines. We gaze headlong into a kind of beauty and terror; terror of systems that charm information out of us, and beauty, because of their design, their efficiency, and the ease with which we give them our fractured souls.

Where are we now? As I write this, at least 4.57 billion people use the internet every day, sending an estimated 23 billion text messages and 293 million emails, making 154,200 Skype calls and 1.6 billion Tinder swipes per day. At present, it's estimated that we produce 2.5 quintillion bytes of data every day. Over the years 2019 and 2020, we created 90 per cent of the data in the world, and this rate of production will increase, year upon year, for the foreseeable future.[17]

I can't describe what 2.5 quintillion bytes of data look like; like the internet itself, it is simply too massive a figure for my brain to make sense of. It strikes me that we have, with the internet, a kind of Prometheus scenario in reverse; even though we created it, the internet has stolen our fire. Perhaps it has even become a god.

We live, and have lived, in a time of acceleration. Start-ups come and go, software proliferates, hardware grows smaller, and our behaviour shifts in response. Technology attempts to tell its history and our own, rewriting the past with each new iteration. It filters and deletes. It updates and erases its tracks. It is constantly rushing onward, forgetting the past, like a figure in a side-scrolling video game.[18]

That's the problem with Timelines; they're untrustworthy. They have little to do with messy human life.

Can writing compete with data, in capturing the human soul? Do you know me better after reading this essay, which took 177 pages of notes, 6,280 words, 37,212 characters, 13 uses of the word 'data' and 13 uses of the word 'self' to compose?

Try as I might to resist its influence, technology permeates my relationships, my work and my sense of self. It has shaped my life's story, and its work will never be done, not until, I believe, it has mapped my soul. In writing, then, I must attempt to convey my soul first.

PART 1
Internet Weirdo

The Night Gym

IT'S 2 A.M., AND I AM coming up on my 700th calorie burned. I'm on a stationary bike that rattles, sweating from the hoodie I've forgotten to take off. At 666 calories I pause and take a picture of the screen with my phone: workout of the beast. As well as displaying my distance cycled and calories burned, the screen offers reassuring messages. *Remember to stay hydrated*, says the bike. *Take long, smooth strokes.* It tells me, *Remember: this is your ride.*

I joined the low-cost supergym in the autumn of 2015, to flatter myself into thinking I was the type of person who went to the gym. It was also a bid to impose order on my life. I didn't know if I could even afford to live in Dublin, the city I'd grown up in, but which was rapidly transforming into one meant solely for the rich. I was freelancing, writing mainly about technology, and while I was happy with this work I harboured a fascination with people who got up every day to go to an office. The idea of predictably sleeping, waking, dressing and seeing other human beings seemed enviable compared to my own unstructured existence, with its mood swings and deadlines and multiple breakfasts.

Looking through my wardrobe of freelance slob-wear, I realised I had become the very thing I wrote about: an internet weirdo, a basement dweller, one acquainted with the night Twitter.

When I first joined the gym I went there in the morning, like a normal person. I figured if I could get up every morning and work out, I'd set myself up for a productive day. Membership cost just €32 per month, including access to classes with names like 'Rockin Rebound', 'Ass and Abs', 'Battlebells' and 'Fight Club'. They promised twenty-four-hour opening, every day of the year except Christmas.

For a while I was entertained by it. I liked it for people-watching, studying the coders and lawyers and marketers. I began to harbour dreams of meeting a heartless libertarian tech millionaire and marrying him. I bought discount designer workout wear in TK Maxx. I bought a phone holder from Tiger that looked like a black Velcro tourniquet. I shopped around for a gym mascara. The gym occupied the fringes of the corporate world, one I could infiltrate and observe. To join its denizens here struck me as a common test of will and dedication: could I be as hardworking, as employable, as they were?

At 7 a.m., I attended a spin class held in darkness with ultraviolet disco lights. The promise was that you would burn 700 calories, with an arm workout built in alongside the frantic pedalling. At intervals we had to stand up on the bikes and manipulate the handlebars through left and right turns, an awkward waltz with a partner that was lifeless and heavy, and fixed to the ground. Meanwhile the sound system blasted a track by David Guetta and Kid Cudi, with lyrics about late nights and hedonism. After the class I took a shower and went home, and fell instantly back to sleep.

I began to overthink the gym, worrying about the correct amount of time to use machines, or how to behave in the locker room, or how to dress. I was unsettled by the degree to which some people dressed up for the gym; I counted around me the number of tiny lower-back logos for Lululemon, the activewear brand which prints quotes from *Atlas Shrugged* on its shopping bags, and which charges over €100 for yoga pants. Around this time, Beyoncé's line of athleisure wear, Ivy Park, debuted at Topshop. I went and looked at it, thinking this was how people are meant to dress for the gym, but I just didn't understand. I wondered if this made me the 'Becky' discussed in online think-pieces: Beyoncé's suburban white nemesis, a girl too basic to pull off high-cut bodysuits and logo leggings.

My problem was not gym clothing. My problem was too much watching, and proximity to people who, I felt, were achieving more than me. The anxiety I experienced in standing and waiting for a treadmill was greater even than that produced by running on one while being watched by the next person in line. Worst of all was when someone using the treadmill or stationary bike before me failed to press the 'stop' button when they disembarked, leaving behind a screen displaying their miles and calories burned as a challenge for me to match.

Throughout the spring of 2016 I grew volatile and fearful; I took on more work, saw my friends less, and felt myself retreating from the world outside. The gym no longer offered respite from these feelings; I entered a state of paranoia, watching the gym-goers and wondering if they were watching me. I began to fear the judgement of women in the changing room for being too naked, or not naked enough. I was becoming scared of other people: the more I went to the gym, the greater my conviction was that they and I were not the same. I measured myself against these 'normal' people and failed. I felt like an alien, using weight machines to work on my human suit.

The day I finally gave up going to the gym in the daytime was the day of Dip Belt Man. He was tall and broad, and had a *Dumb and Dumber* haircut, a vest top and ill-fitting shorts. He was wearing a thick black belt, which gave him the appearance of a 1990s WWF superstar. I noticed him because he'd been standing in the same place for fifteen minutes. I knew this because I had been on the bike that long, wondering who the fuck this guy was, who was staring and making me nervous.

As the podcast about serial killers I was listening to wound up, Dip Belt Man finally started to move. I watched as he took one of the largest weights from the rack in the corner, unclasped the belt and hooked it onto his waist. Slowly he swung it around to the front, until it dangled from his crotch like a cast-iron codpiece. Then he went back to staring and standing in the same place.

Bodybuilders use this method for weighted pull-ups, when lifting the mass of their bodies isn't enough. The belts resemble devices from a medieval dungeon, fashioned from chains and neoprene. But it wasn't the belt that disturbed me: it was Dip Belt Man's eerie stasis, and his utter lack of self-consciousness.

I concluded that day that the Day Gym – as I later came to think of it – was for people who were prepared to be put on display. It had not escaped my attention that the word 'gym' comes from the Greek γυμνάζειν, a verb meaning 'to train naked'. I wanted to work out among basement dwellers and freelancers, like me. I wanted to be an odd fish among clams.

I knew that the gym was open twenty-four hours, and that unlike the rest of the working world my life had no rhythm, no obligation to wake in the morning and sleep at night, when I could have the gym almost entirely to myself. So, I would give the Night Gym a try.

On my first trip to the Night Gym, I arrived at midnight and noticed a nu-metal and hip-hop soundtrack: Eminem, Limp Bizkit, Crazy Town, Red Hot Chili Peppers. Nearing 1 a.m., the music stopped entirely, along with the air conditioning, and the screen that normally played old Popeye and Mickey Mouse cartoons went dark. The lights dimmed on those of us who remained: a solitary weightlifter, a woman on the stair climber and a man sprinting on the treadmill, set to its fastest setting.

I liked the darkness: it felt like cover. I returned every night of that week and the week after that, staying later with every visit.

My favourite part of going to the Night Gym was going to the Night Gym – which is to say, the walk from my house. The air was cool; the lights of Custom House Quay glimmered on the river. The McCann FitzGerald building, a giant square made of glowing neon blue cubes, seemed safer by night when the lawyers were gone. If I timed it right, the streets were empty of cars and late-night drunk people, too. On the streets I caught glimpses of a nocturnal economy: the flight crews pulling suitcases on wheels, and drowsy staff pouring out of the bars and restaurants around CHQ.

At the Night Gym, I began to notice the regulars. There was the late-middle-aged man doing weight machines to build up his skinny arms. There was a group of young Asian guys who I imagined worked together somewhere on a late shift. One weekend night there was Drunk Guy, wearing a tight shirt and dad jeans, bootcut, over leather loafers, one of which fell off his foot as he crossed the floor. His eyes were half open. His feet dragged and he dropped his wallet as he made his way to the machines. In one hand he held a bottle of Budweiser, which he set down before beginning a set of chest-presses. Nobody stopped Drunk Guy, and everyone watched him. What drove him from the night that he was dressed for, and into the gym? Why were the rest of us here sober? Deep in the lizard brain, would we all remember how to use a chest-press machine, even during a blackout?

*

The year 1995 marked the first mention of Ireland as a tech hub; in an interview with the San Jose *Mercury News*, a Harvard business professor named Rosabeth Moss Kanter described us as 'the Silicon Bog'.[1] By 1998, *Wired* magazine was calling us the Silicon Isle, and by the early years of the new millennium we'd established the Silicon Docks, a term still used today to describe Dublin's tech district. Google played a role in its creation; in 2004, after courtship by Ireland's Industrial Development Authority (IDA), they set up their EMEA headquarters on Barrow Street, and eventually spent €250 million on four office buildings.[2] Rents in the neighbourhood climbed, other tech companies arrived, and gradually the Docklands transformed from a neglected inner-city suburb into a Silicon Valley outpost.

Years ago I did some freelance writing work for an Irish venture capital firm, helping to create a guide to Dublin and its tech scene for visiting entrepreneurs. I remember it featured a series of maps of Dublin; the first, dated 'Pre-2000', featured Oracle, Microsoft, Dell, IBM, HP, Intel and Sun Microsystems, dotted mostly around the outer suburbs in campuses and business parks. The next map, for the early 2000s, saw Facebook, Google, eBay, PayPal, Amazon and Salesforce set up in Dublin too. Finally, between 2010 and 2012, there was a rush of new arrivals: Twitter, LinkedIn, Dropbox, Etsy, Airbnb, Zynga. Almost every well-known name in tech had established a base in Dublin, most often in the Docklands.

I worked on and off within this self-declared 'tech ecosystem', mostly writing blog posts and web copy, and occasionally running social media accounts. I was expected to downplay the low Irish corporate tax rate of 12.5 per cent, and emphasise our 'high-quality local talent' instead. Occasionally I'd go to meet-ups and start-up launches. There was always a free bar with a corporate sponsor, and speeches extolling the virtues of entrepreneurialism, of burning through funding and, ultimately, of failure. 'Fail again. Fail better', I heard repeated on every occasion. I had just graduated from an MPhil in literature. I fantasised that one day the Beckett estate would catch on to these start-up people, and sue them for everything they had.

I often wondered if this was a bubble. The nation was marketed abroad as 'Innovation Ireland'; a land of *céad míle fáilte* filled with English-speaking, overly qualified potential customer service agents. Very little tech IP was created in Ireland; a large number of the jobs were in marketing, customer service and content moderation. Our time zone, Greenwich Mean Time, was eight hours ahead of Silicon Valley's, and suited the idea of 24/7 business. One IDA print ad published in business magazines read 'Facebook found a space for people who think in a certain way. It's called Ireland.'

The new wave of tech companies in Dublin were assigned a category, 'born on the internet', meaning they relied on ubiquitous connectivity to remain in business. Their founders, programmers and customers were very likely born on the internet too. As in California, the Silicon Docklands fetishised youth; throughout my twenties, I watched as casually tech-minded acquaintances founded start-ups and attracted funding with little more than a hoodie, a pitch deck and a LinkedIn account as qualification. Money was thrown around with a fervour not seen since the Celtic Tiger, but under the guise of something less frivolous: an investment in Ireland's future.

I remember being brought on a tour of a new start-up accelerator in the Docklands in order to write about it; a former wine cellar, dating back to the Georgian era, had been turned into an underground business hub. Dim, cavernous halls the size of football pitches lay dormant under the Silicon Docklands. Technology had colonised this space; already it had covered the surrounding area, reaching into the sky with ambitious, oddly shaped towers. Now it was clawing downward, digging up history, and finding it fertile ground for 'growing' more businesses.

Eventually I stopped doing marketing, and began to find work as a freelance journalist. I mostly wrote for international websites, earning a small and precarious living. I was never really sure where I fit; I had a middle-class South Dublin accent, a good education and very little money. A steady stream of articles was published about the rental crisis, about mortgages and zero-hour contracts and how young people could no longer live in Dublin. Money felt like a joke to me; I would put in long hours, staying up all night researching stories, and in the daytime I carried my work around in my pocket, on my phone, and with my laptop in my bag in case I suddenly needed to work on edits. I didn't have work hours – I simply worked until it was done, and then took on more work – and often the pieces I wrote paid a set fee of €100. Mostly I lived on savings from years before, when I worked on commission selling make-up in a department store, earning considerably better money.

The shared house I rented a room in was on a street just off North Strand, a historically neglected area of north inner-city Dublin. During the Second World War, in which Ireland was neutral, it was famously bombed by the Luftwaffe, but these days drug addiction was a far more significant problem, with a memorial nearby on Sean McDermott Street featuring a 'flame of hope' in memory of local people who had died due to heroin addiction. Down the road in Ballybough, a feud was accelerating between criminals allied to a local figure known as 'the Monk' and the Kinahan gang, an international drug cartel.

From these streets you could just about see the effects of tech-driven urban regeneration, a ten-minute walk away. Towers and odd neon geometries lit the night sky; the blue bands of light on the Convention Centre; the uneasy curve of the Samuel Beckett Bridge. I pictured the trajectory of a tech executive, arriving at Dublin Airport – doubtless at the new, shiny Terminal 2 rather than Terminal 1, which was run-down and only used by Ryanair.

They would be driven along the Liffey to a newly built hotel in the Docklands, before meetings in one of those towers, from which they would see other shiny new start-up headquarters and hotels. It was a version of Dublin designed to be seen from a boardroom: tall buildings, surrounded by other tall buildings, air-conditioned and clinical and gleaming.

I would walk out the door, and then under an old train bridge where I occasionally saw people passed out with syringes in their arms, past the rumoured IRA pub and the defunct mattress shop, then past the bus station and into the Docklands, moving from former tenements into new architectural complexes more suited to Dubai. Online one night, I read about Silicon Valley's 'hourglass economy': the concentration of very rich and very poor people, with few residents falling in between. Something similar was happening in Dublin; house prices and rents had skyrocketed, as had the number of homeless people.

One night I walked in the dark along North Strand, and paused at the railway bridge to watch a group of teenage boys gathered around a bonfire in the trainyard. There were scraps of wood, furniture and old tyres burning in front of them. Behind them, across the water, was East Point Business Park, a campus I had once visited for work. They stood in a loose circle, staring into the flames, and facing an enclave of a new and stratified Ireland; Astroturf greens, car parks, bright cafeterias and the offices of Cisco, Citrix, Oracle and Enterprise Ireland. I watched them a while, then continued on my way to the Night Gym.

If you're running, then you must be running somewhere. At the Night Gym, I rarely stopped to question where.

In 'Against Exercise', an essay written in 2004, Mark Greif writes of how the numbers on the treadmill's screen become talismanic objects of faith. To the breathless runner, these figures are 'even more preoccupying than your thoughts or dreams. You discover what high numbers you can become, and how immortal. For you, high roller, will live for ever. You are eternally maintained.'[3] I'm not quite sure how I arrived at the goal of 700 calories to burn per day. Lacking any idea of what a 'good' number of calories burned might be (but certain that there must be one), 700 became my mental crutch, a simple, numerical solution to a lifetime of disordered eating: accomplish this, and spend the rest of the day not feeling bad.

Like work, like the scrolling social media feed, like data centres and disk storage and money, activity in the gym is potentially limitless. The virtual track could go on for ever, as long as you're willing to run it.

The treadmill was invented in England in 1818 as the 'everlasting staircase', for the purpose of convict reform.[4] A prison guard, James Hardy, wrote that it was the treadmill's 'monotonous steadiness, not its severity, which constitutes its terror'.[5] Men were lined up in chains and made to walk a giant wheel, enacting their sentences as wasted time spent pacing towards eternity.

After eight decades of use in prisons, the treadmill was banned as 'excessively cruel'. Later it made a commercial comeback, first for dogs whose owners were too lazy to walk them. Then in the 1960s, a treadmill marketed as the 'Pacemaster 600' began to appear in people's homes, for voluntary human exercise.

Today it's normal to allow machines to coach us and even hurt us. We watch videos of 'treadmill fails' on YouTube. In 2016, Apple products were marketed with a video of Taylor Swift falling off a treadmill while rapping along to Drake, and an estimated 20,000 people show up in emergency rooms each year due to treadmill accidents.

Online one night, before leaving for the gym, I found a forum called 'OfficeWalkers.com', where users trade tips on how to build treadmill desks at home. The tagline is 'Working at 100 calories

Per Hour' (I wish I could tell them the sedentary body already burns 65 calories). One picture on Office Walkers stands out, of a man standing inside a gigantic wooden hamster wheel. He has coated its insides with grip tape, and there's a laptop desk across its frame. An eleven-second video is hypnotic: in his garage, he has constructed his own solitary endless staircase.

Fitness culture loves to discuss 'evopsych', also known as evolutionary psychology. On websites, bodybuilding forums, and in the copy you might read on a box of protein bars, the consensus is that our bodies have not yet moved on from the Stone Age. We yearn for the cave, the physical challenges of a life before civilisation. Framed as such, exercise becomes an act of longing for the past.

One problem with putting this belief into action is that the world these fitness enthusiasts inhabit is not hospitable to cavemen. Cities don't offer much in the way of savannah to run across, or dire wolves and mammoths to battle with, or oversized boulders to push up hills. Still, the true believer finds ways to adapt, to reconnect with their 'natural' self.

In London, and New York, and Shanghai, and other parts of the world that have skyscraper districts, the city is repurposed in a practice called tower running. Every so often a group of elite athletes will sprint up and down the stairs of a tall building for money. Past races have taken place at Colombia's Colpatria Tower (980 steps), the Empire State Building (1,576), the Menara Tower of Kuala Lumpur (1,850) and the Taipei 101 (2,046). There's even an organisation called the Vertical World Circuit, which stages races yearly. Its logo depicts a small man in shorts running towards a cluster of skyscrapers, drawn in Art Deco style in black and gold. It resembles the insignia of a supervillain, or the cover of a book by Ayn Rand.

You can watch tower races on YouTube. They usually begin

outside the building's lobby, in a conventional race track set-up, before the runners dash inside. Then the camera cuts to a windowless staircase, the kind people who work in skyscrapers rarely pay attention to unless their lift is broken. The contestants lurch past in turn, hurtling skyward, gripping the handrails and taking steps two at a time.

As one tower runner disappears out of the frame, another one surfaces one storey behind him. They have numbers pinned to their shirts and some are barefoot, as though bound on a hellish pilgrimage into the sky (others wear Vibram 'FiveFingers' shoes, which mimic the experience of running barefoot instead).

Finally the camera cuts to the end of the race, an office lobby on the top floor suddenly invaded by men in Lycra. At the finish line they fall to their knees on the polished tiles, reeling with euphoria and the horror of so many stairs.

In Dublin, the people who go to the low-cost supergym work in the Facebook HQ on Grand Canal Square, or the Iconic SOBO Works co-working space around the corner (in 2015 the area was branded as an acronym for 'South of Beckett O'Casey', referring to the bridges on the Liffey, in a campaign widely mocked by the Irish public),[6] or at MindGeek, the 'web design and SEO' company which is actually among the largest online porn providers in the world (Pornhub, RedTube, Reality Kings and Brazzers all fall within its remit).

The Silicon Docklands are marketed as a success story arising from Ireland's ambition, its industriousness and national genius, with the low corporate tax rate offered as an afterthought. Ads for office space in the Docklands describe the area as 'bustling', but this is wrong: the Docklands do not 'bustle'. They only fill up in the mornings with people who walk in straight lines, and empty out again every evening.

Silicon Valley culture is preoccupied with youth, and health, and physical purity, leading tech CEOs to embrace trends including intermittent fasting, self-monitoring with apps and devices, the consumption of nootropic drugs, all-meat diets, and, if rumours about one high-profile venture capitalist are to be believed, the transfusion of blood sourced from young people into older, wealthier bodies. The tech industry has also long associated itself with biohacking, a distinctly male reinvention of wellness culture which branches into fitness and diet, 'cognitive enhancement' and life extension. Falling at an intersection between culture, technology and the body, biohacking promotes a vision of health as something that can be engineered, controlled and quantified with tech products and devices.

Jack Dorsey, co-founder of Twitter and founder of Square, has attracted both scrutiny and praise for his spartan lifestyle experiments, which include daily fasting, five-mile walks, ice baths and the use of a near-infrared sauna installed in his garage. In the morning he consumes 'salt juice', a mixture of pink Himalayan salt, lemon and water, which is also dispensed worldwide to Twitter employees. He also goes whole weekends without eating. Dorsey told the Ben Greenfield Fitness podcast, of his morning ice baths, 'I feel like if I can will myself to do that thing that seems so small but hurts so much, I can do nearly anything.'[7]

It's clear to me that these practices are about pain, and endurance, and, I'm quite certain, guilt, at the wealth Dorsey has acquired from running one of the most controversial online platforms in the world. The tech industry is characterised by extremes, driving its leaders to cultivate extreme bodies. It also strikes me that Dorsey is suffering from a very masculine, very Silicon Valley version of an eating disorder, one that has evolved in response to a life, and a legacy, that has spun out of control. But I'm not a doctor, and I can't diagnose Dorsey; I'm only a bystander with a similarly uneasy relationship to technology, the body and control.

Dorsey is by no means alone in his habits. The tech industry has staked a claim on the body as a battleground, a space for proving the triumph of innovation over medical science and physical frailty. Inventor and futurist Ray Kurzweil is said to take 180 to 210 vitamins and supplements per day; one 2008 interview with *Wired* saw him spending one day of each week in a medical clinic, receiving injections of phosphatidylcholine in a bid to 'rejuvenate all his body's tissues'.[8] Kurzweil's hope is to live for ever, or, at least, long enough for technology to reach a point of Singularity, through which he will transcend his body completely. Doug Evans, founder of the infamous failed juice start-up Juicero, is now an advocate for 'raw water' instead, sold by a company called Live Water. He famously fasted on the unprocessed, unclean bottled 'product' for ten days during Burning Man.[9] Then there are the less hard-line, more accessible biohacks; the nootropic supplements, the macrodosed vitamins and the microdosed hallucinogens, which still, by contrast, make going to the gym at night seem rather basic.

Jack Dorsey's daily routine, and the media's fixation on Jack Dorsey's routine, speak to a cult of the tech founder which can be traced back to the time of Steve Jobs, whose habits included wearing one of 100 identical black Issey Miyake turtleneck jumpers and blue jeans every day, a system he developed to prevent 'decision fatigue'. Jobs's uniform inspired younger CEOs like Mark Zuckerberg (hoodies, T-shirts and Adidas flip-flops) and Elizabeth Holmes (a black polo neck – Holmes made herself a female carbon copy of Jobs), and kickstarted an ongoing trend among tech founders for clothing as an aesthetic expression of focus and self-discipline.

Tech culture esteems humans who act like machines; it prizes repetition above spontaneity. It's about luxurious abstention (restraint, but only in the context of vast wealth), and a need for control, a hyper-capitalist take on the anorexic's saintly self-abjection. Much like Apple hardware, where the blood, sweat and silicon which go into manufacturing tech products are concealed beneath a clean, minimalist casing, the Silicon Valley founder conceals their will to profit, their drive for data and expansion, behind a veneer of simplicity; they perfect the look of someone who has transcended their own wealth.

The lifestyles of tech founders filter down into company culture, translating into Paleo lunch options and mornings of company yoga. Desk hours are counteracted with hours in the gym, which are woven into the routine of working life, and gyms appear in built-up urban areas, nestling among office blocks, occupying basements or space above shops. For someone whose world has shut down into a circuit of office, commute and home, the gym provides the illusion of space: it offers a limitless supply of terrain to run, cycle or sprint across, magically contained in a single location.

At the low-cost supergym I ride a bike with a screen attached, where you can choose from a range of 'virtual tours'. These programmes are designed to keep you mentally engaged, mimicking beaches in California, rural French cycling routes and mountain ranges in Peru. You race only against yourself: a tutorial explains that 'ghost riders' will appear on the track to represent your tours from previous sessions, so that when you beat your record, you pass through your own ghost on the track, watching it dissipate into pixelated air.

If Silicon Valley were a country, it would be among the wealthiest in the world.[10] It's the part of the world that brought us the unicorn, blitzscaling and vaporware, a land of valuations that often have no bearing on reality. Among these young companies and their founders, there is a belief in moving fast and breaking things – breaking people – as part of a quest for near-limitless growth. Like Ray Kurzweil's belief that he can live for ever, the average tech CEO does not envision an endpoint for their product beyond ubiquity. Once that's achieved, a new kind of limitlessness becomes the goal; like Google, Amazon and Facebook, they can diversify into new fields and monopolise them too.

In 2011, investor and entrepreneur Marc Andreessen wrote that 'software is eating the world', arguing that tech companies were 'poised to take over large swathes of the economy'.[11] Technology will remake the world and monetise its behaviours, and this sense of infinite growth is coupled with a mania for quantification. As if in response to the grandiose nature of their never-ending mission, companies track the productivity of their workers with the same technologies they sell.

One tech company with an office in Dublin, 'experience management' organisation Qualtrics, created a software called Odo that monitors the workplace. A *Fast Company* profile of Qualtrics from 2016 reads 'The sharing is meant to encourage others to get a glimpse of the top performers; it encourages a sort of competitiveness mixed with community.'[12] Cameras are positioned in every room, over the desk of every employee, while a company-wide dashboard – the 'odometer' of the name – records everyone's weekly tasks and whether or not they accomplished them.

During the Cold War, defence companies in Northern California began to attract workers with what they called the 'sci-tech personality': single-minded, socially awkward people characterised by a devotion to their work to the exclusion of all other things.[13] This tendency – a robotic coldness coupled with 'passion' for work, gave way to tech's latter-day mantra of 'do what you love', a line most often attributed to Steve Jobs because of his Stanford commencement address in 2005, in which he said 'The only way to do great work is to love what you do.'[14]

In tech, work is relabelled as love; I've noticed that tech founders, more than any other kind of boss, tend to assume in interviews that other people are as personally invested in their companies as they are. This applies especially to millennial employees; if they're not appropriately 'passionate' about their jobs, then they're dismissed as incompetent and lazy.

The problem lies in quantifying love, and, beyond that, the meaning people derive from their work. The crudest measure is also the most common; workplaces measure their employees' 'passion' through hours spent in the office, which is often equipped with free drinks and food, nap rooms, foosball tables and other Google-y clichés to encourage them to make themselves at home. In tech, the meaning of work itself dissolves; in an article for *Jacobin* published in 2014, Miya Tokumitsu wrote, 'In ignoring most work and reclassifying the rest as love, DWYL [do what you love] may be the most elegant anti-worker ideology around. Why should workers assemble and assert their class interests if there's no such thing as work?'

'Do what you love' overlaps with the mission of tech companies themselves: limitless growth, at any cost. For years, Mark Zuckerberg would end Friday all-hands meetings by shouting 'Domination!' Later, in 2018, a memo by Facebook exec Andrew 'Boz' Bosworth was leaked, titled 'The Ugly', in which he wrote:

> The ugly truth is that we believe in connecting people so deeply that anything that allows us to connect more people more often is *de facto* good ... That's why all the work we do in growth is justified. All the questionable contact importing practices. All the subtle language that helps people stay searchable by friends. All of the work we do to bring more communication in. The work we will likely have to do in China some day. All of it ... The best products don't win. The ones everyone use [sic] win.[15]

Phrased this way, working in tech becomes a confrontation with limitlessness. It means meeting a demand for endless expansion, endless data and endless 'passion' in the working day. This climate estranges the mind from the body; sometimes it negates the body entirely, or treats it as an automaton for the mind to control. A three-day water fast becomes a way of quantifying time, of controlling it and making it apparently worthwhile, at a cost to the body's needs. It takes the logic of the coding sprint, the all-night energy drink-fuelled work binge, and applies it to the body.

This extreme behaviour appeals to a certain personality type, but it also usually means setting oneself up for failure. Measurability, self-monitoring, biohacking, growth hacking – any kind of hacking other than that done to a machine – sells the promise of control to people who, on some level at least, know they're working in an industry where they'll soon be obsolete, or, at best, overwhelmed by their own success. Speaking at TED's 2019 conference, the theme of which was 'Bigger Than Us', Jack Dorsey questioned his own decision, made over a decade before, to build quantification into Twitter as a central feature:

> If I had to start the service again, I would not emphasize the follower count as much. I would not emphasize the 'like' count as much. I don't think I would even create 'like' in the first place, because it doesn't actually push what we believe now to be the most important thing, which is healthy contribution back to the network and conversation to the network, participation within conversation, learning something from the conversation.[16]

The numbers haunt us; there's no prize, and no end, because no one ever wins.

The word 'balance' appears frequently in fitness literature. A 'healthy balance' is desirable, between work and home, stasis and activity, desk and gym. Balance, for me, has often been an unhealthy juggling of extremes: limitless work and limitless workouts. We recreate the dazed, exhausted, end-of-day feeling we get from a day of work, but this time in motion, and purely for ourselves, at the gym.

Over the span of my life, I have learned precious little about balance. From around the age of eleven I was anorexic, refusing to talk to anyone about this problem even when it was impossible to conceal. Later I became bulimic, in my first year of college, and the secrecy and ritual of the thing very quickly became addictive. It escalated to a point that, looking back now, I find rather baffling; sometimes I'd throw up over thirty times in one day. I don't even know how I found the time for it. I kept that secret, too, until a series of emergency root canals scared me out of the habit.

After they have recovered, people with issues like mine often find new and socially acceptable ways to hurt themselves. By the time I started going to the Night Gym I was eating enough food to be relatively healthy, but the gym became an outlet for behaviours – the same old neurosis, self-loathing and secrecy – which had never really gone away in the first place. It wasn't just about avoiding working out in the daytime. It was about taking something mundane – as tower runners do with staircases, and as I had once done with food – and making it extreme and potentially dangerous. The gym was an apparently healthy activity, rendered unwholesome by practising it in the dark.

Years ago, when I started working as a freelance writer, one of my editors advised me to treat it like a real job. He told me, 'Get up in the morning, put on clothes – real clothes, not sweatpants – and go to a cafe with your laptop.' He advised me that the Starbucks on Lower Leeson Street was a good place to work because there was an upstairs with plugs in the walls, and they didn't mind if you stayed for hours with only a single coffee. 'The structure will be good for you,' he said.

Fitness instructors warn their clients of 'plateaus', of reaching the point where progress ceases and you enter an unproductive cycle. After six months of going to the gym at night, something similar had happened to me, not only in terms of exercise but also with my life in general. What had begun with the intention for self-improvement had given way to something less wholesome.

Throughout the summer of 2016, as slowly, unwittingly, I edged towards a breakdown, I clung to this unusual routine as proof of my sanity. On the average day I woke up at six, read the news and watched YouTube videos. By eight I'd be seated in a cafe somewhere in the Docklands, watching people with actual jobs go to work, and apparently 'working' on my laptop. By mid-afternoon I'd go home and fall asleep, or wander around charity shops, or smoke weed in the overgrown back yard of the house I shared. Then I'd work some more, past midnight. Finally I'd change into leggings and a hoodie, and go to the gym until 3 a.m. or 4 a.m.

This system I had devised guaranteed enough work to pay rent, and a modest budget for everything else. But it also drew me further and further away from contact with other humans, leaving only the company of machines. Gradually I went out less and less, apart from the gym. As in the gym, where the proof of a good workout is an aching body, I told myself that each cancelled invitation, each ignored event or unanswered message was proof that I was working hard. Loneliness had become addictive; it had come to define me.

There's a revelatory phase that young people pass through when they first realise that they can have adult lives away from their parents. It takes a while to find a balance; often for a time you end up drinking too much, or taking too many drugs, or – a milder excess – failing to learn to cook, and living on takeaway pizza. This lack of moderation goes both ways; it's also easy to work too hard, and in unnecessarily exhausting ways, especially if you're alone and lacking a template of what a working life should be.

At twenty-six, that period should have been long behind me. But my life during this time was characterised by a persistent feeling of not being enough. I was that person who sends emails in the middle of the night. I would sit upright in bed with my laptop overheating across my knees, wondering which part of my body would atrophy first, before going to the gym to furiously counteract the hours of stasis. I ate a narrow, caffeine-heavy diet, mainly involving Special K cereal, microwaved vegetables and cans of Monster Energy Rehab, a guarana–iced tea hybrid drink which generated a slight ringing in my ears. The copy on the side of the can, apparently written by Monster's owner, provided welcome distraction: 'While chillin' at the Vegas Rehab® pool party, contemplating a cure for cottonmouth, admiring the flesh parade, and pondering the wisdom of doubling down when the dealer shows a face card, it HIT ME! We need a new drink.'

The combination of taurine, caffeine, Panax ginseng root and lemon flavour was aimed at athletes, late-night workers, and, it could be inferred, people in need of clinical treatment for addiction ('Rehab the Beast!').

Summer came, then began to fade, and slowly I felt myself fall off the edge of the planet. I kept working, taking on more deadlines and forgetting to send invoices. My days narrowed down into a cycle of exercise, sleep, typing and scrolling, watching online society unravel and feeling myself grow incrementally more numb.

When I wasn't at the gym I would spend long periods sitting at my desk, staring at the emails in my inbox, but unable to open any of them. At night I would sit in the dark, face lit by the screen's soft light, examining the timelines of journalists and writers and strangers, whose tweets about politics and brunch pictures had attracted more Likes and shares than my account had since I first signed up. I felt the life drain from my body through my eyeballs, my fingers weakening as I typed. I hated and envied other people their confidence. I felt guilty for hating them, but then rationalised that I hated myself even more.

I read about burnout. I read about the shut-in economy, about zero-hour workers and a generation of young people leaving Ireland to find work abroad. I loved writing, and felt fortunate to be able to do it as my work, but I couldn't enjoy it. Life came to be defined by a strange, contradictory sense of limitlessness and restriction. The goalposts kept moving; I worked in the name of some far-off point when I'd be able to take a break from work, and by night I ran, so that at some point I wouldn't need to keep running any more.

It is possible to practise habits of self-improvement, and at the same time continue to long for your own slow destruction. Research on the brains of people suffering from eating disorders shows a decreased sensitivity to rewards; fasting, especially, skews your ability to think in terms of long-term goals, or to imagine the future (in interviews, Jack Dorsey describes this effect in psychedelic terms, as a hallucinatory 'new dimension' where time slows down).[17] Much like anorexia, where you quantify and reduce but ultimately will never be thin enough, I took on more work but there was no fixed goal to achieve, no point at which I would allow myself to relax.

Eating disorders are often depicted as being about food and dress size. We're given a picture of a girl looking into a mirror and seeing herself as fat. The reality is something more general, which goes far beyond eating; a thousand tiny brutalities, diffused into everyday life so as to be almost invisible. For the sufferer, life becomes an endurance challenge, and mundane tasks like sleeping or socialising, or allowing oneself to relax, are drafted into the game.

By the end of the summer of 2016 I realised I had a problem, not just with my workout habits but with the order of my life in general. Messages went unanswered, calls unreturned. My depression and paranoia spiralled, till finally I took the overdose, indisputable proof that my life required urgent change.

I told my editors I needed some time off. I moved back in with my parents, and several weeks later I entered a treatment programme for depressed people. This meant, for the first time, acknowledging my limits; I said no to work, signed out of Twitter, and gradually realigned with the world, sleeping at night and working during the day.

Lately I don't go to the Night Gym. I used to think of Dublin as provincial, and envy those who lived in twenty-four-hour cities. Now, I feel a little sad at the prospect of others becoming like myself: joining the ranks of the people who run instead of sleeping. In the Silicon Docklands, cranes loom in the distance beyond the neon towers. Someone is scrolling, someone is working late, and someone, somewhere, is running on a treadmill. Night in Dublin is day in California, and there's always the chance – the open invitation – to rejoin the race.

Bland God: Notes on Mark Zuckerberg

NOT LONG AGO, in a charity shop on Capel Street, I found an expensive hoodie. Not expensive in terms of its current price – the Goodwill Thrift Shop was asking for €4 – but its original one. The hoodie bore no outward sign of branding, but its design was strategically luxurious, sewn from the softest cotton and dyed to a muted, achromatic grey. It had been designed to include the kind of thoughtful details you only notice on second glance, when you realise the person wearing it is a little wealthier than they first appeared.

Second-hand shopping is unpredictable. For every useless, possibly cursed item gladly donated by its former owner – the monkey's paw, the dybbuk box, or the shoes that make your ankles bleed – there's a miracle find, that ludicrous, beautiful thing which has found its way to the charity shop by chance. I might once have laughed at the idea of owning a designer hoodie, but, touching its fabric, I felt myself converted. I pulled the zipper, looked inside to read the label, and noticed spidery text across the back on the inside lining: 'Making the world more open and connected.'

It was a hoodie given to Facebook employees, likely donated

to the shop by someone who had recently left the company. If walking around wearing Facebook's mission statement was an uncomfortable thought, then walking around wearing it secretly, printed inside my clothing, made me feel even more squeamish. I put the hoodie back on the rail and moved on.

The hoodie I found was identical to those worn by Mark Zuckerberg at public appearances – an anonymous item that, nonetheless, feels essential to his personal brand. In 2017 a group of writers took Zuckerbergian semiotics, including the hoodies the Facebook founder wears for public appearances, as the subject of a collection of essays. Volume One of *The California Review of Images and Mark Zuckerberg* was edited by Tim Hwang, a fellow at a research institute in New York called Data & Society.

Each essay in the collection addresses an image posted to Mark Zuckerberg's Facebook page, pre-approved by the Facebook founder and his team before being published to over 100 million followers. Zuckerberg's carefully stage-managed presence runs through these essays as a thread, prompting questions about mediation and identity – both Zuckerberg's, and our own.

Why study Mark Zuckerberg? Over the years, as his product has grown to encompass millions, then billions of users, he has revealed very little about his personal life. But Mark Zuckerberg's identity is directly linked to that of every Facebook user, because Facebook, his product, is an identity machine. The site feeds on our identities and all the 'monetisable' ad data they generate, even as it encourages us to manufacture new ones. More than any form of online media before it, Facebook has foregrounded the 'personal brand', the fashioning of a public-facing self.

Amid the noise, the emotional and visual clutter of the Facebook experience, Mark Zuckerberg is notable for his absence. While he has made occasional public appearances over the years, including at the European Parliament and the United States Senate, and routinely publishes Facebook posts addressing his followers, his public image remains subdued, carefully curated, and so ordinary as to be inscrutable.

One essay, 'Mark Zuckerberg's Significant Insignificance' by Dilara O'Neil, addresses this theme, using a 2013 Zuckerberg profile photo as its starting point: 'Here is Mark Zuckerberg, our current tech dude of the moment who presents himself neither as a businessman nor a thinker, but as a Normal Guy.' The image, a little grainy, and taken on a now-outdated phone, features the Facebook founder standing against a landscape which might be the surface of a distant planet. Blank, wide-eyed Zuckerberg poses before blank and rolling hills, under the kind of sky artists in Hollywood were once paid to paint as background scenery. His appearance could be called 'youthful', but Zuckerberg has not noticeably aged in the time since the picture was taken. He appears, then, the same as he always has.

This picture might be a selfie. It might have originally included people who were cropped out. It might also be revealing, but it's not; Zuckerberg's expression is placid, impenetrable, generic, like a child's drawing of a human face. O'Neil draws meaning from the landscape instead, implying that Facebook will conquer it:

> Facebook is not just a website like the ones that came before, which stayed stagnant until consumers grew bored and moved on. With its constant updates and algorithmic strategies to infiltrate user's information, it unleashes onto untouched land and like nature, will continually evolve.

'Facebook is not just a website like the ones that came before'; Zuckerberg was born in 1984, part of a generation influenced profoundly by the arrival of computers, and later the internet, in their childhood homes. For this generation, the same one I am part of, the internet is both a tool for self-fashioning, and the cause of an identity crisis.

As I search for metaphors to describe my own altered state – is it evolution? Mutation? Sickness? – in Zuckerberg I see the other side of this change. Here is someone who has acquired power beyond measure, without precedent; the architect of systems more widely adapted, and more intimate with the human psyche than any form of capitalism before (unless, perhaps, we consider religion). Is it possible to find a way to describe him? Is there a place in language for a figure so connected, yet so unknown?

As Facebook's leader, Mark Zuckerberg is defined by the community – the product – he has built. O'Neil compares Zuckerberg to 'MySpace Tom', a friend to everyone but known intimately by very few, and a guiding figure, who added you on arrival at MySpace and led you through the social media underworld. Certainly there are similarities; you get the sense that Zuckerberg would like to be everybody's friend, if only he could make himself more likeable.

But Zuckerberg is not likeable. He has middling-to-negative portrayals of himself to contend with: Kate Losse's memoir of working at Facebook in its early days, *The Boy Kings*, portrays him as naive and more than a little bit entitled, while David Fincher's 2010 biopic *The Social Network* casts him as inept, bordering on sociopathic. In interviews, Zuckerberg comes across as clear-eyed but lacking in self-awareness; another essay in the *Zuckerberg Review*, Mél Hogan's 'Sweaty Zuckerberg and Cool Computing', takes as its subject a video rather than a picture, of a somewhat notorious interview dating from 2010. At the *Wall Street Journal*'s D8, an executive conference promising 'straight-up, unvarnished conversations with the most influential figures in technology', a 26-year-old Zuckerberg is grilled by journalist Kara Swisher. Visibly sweating when asked about Facebook's privacy settings, he refuses to take his hoodie off, and displays the awkwardness that has come to define his public persona.

Beyond the images curated by his team and posted on his Facebook page, what do we know about Mark Zuckerberg, the third-wealthiest person in the world? What does he stand for, apart from the cause of Facebook itself? This question might connect to a larger one: what does Facebook itself stand for? The slogan on the hoodie I found, 'Making the world more open and connected', has since been discarded as a company motto; in 2017, Facebook announced a new and equally innocuous intention to 'give people the power to build community and bring the world closer together'.

The world was 'open' before Facebook, a time when communities existed just fine. It could even be argued, especially in light of the 2016 US election and the role of consulting firm Cambridge Analytica, which misused the data of roughly 87 million Facebook users, that services like Facebook actively narrow our perspectives on the world by showing us only what we want to see.[1] Facebook reinforces the 'filter bubble', a term for the effect of personalisation algorithms coined by Eli Pariser in his 2011 book of the same name; as Pariser wrote, 'a world constructed from the familiar is the world in which there's nothing to learn'.

Facebook's vision of community, much like its vision of friendship, wears thin under scrutiny. 'Community' implies a group of people with beliefs, or interests in common. It suggests a network of care. But what Facebook's users share is that all of them generate data, media and interactions, which earn money for Facebook. What Facebook cares about, meanwhile, is its ability to continue profiting from this business model without taking responsibility for what happens on its platform (as suggested in Andrew Bosworth's memo, 'The Ugly', the consequences take a backseat to 'connection', and potentially infinite growth).

Facebook is far from alone in its approach. Silicon Valley's Big

Five (Google, Apple, Facebook, Amazon, Microsoft) and their 'disruptive' heirs (Uber, Airbnb and others) share a common response when faced with criticism. If a violent, bigoted or otherwise objectionable video appears on YouTube, or a hate group forms on Facebook, or if your Uber driver tries to rape you, it is never the fault of the company. This is because the company does not 'employ' the driver, or 'endorse' any of the content its users create. They may not even be aware of the issue, because the platform is too big. Lacking the grassroots values of early internet culture (where users were relied on to police themselves, but retained some right to ownership of their content in return), platform capitalism blames the user when anything goes wrong, mocking them for being misguided enough to trust the service in the first place.

Facebook is not a community, but a platform; it is a blank, hungry canvas which swallows up and monetises its users. While Facebook's reps pressure media outlets into paying for more 'native advertising', the company denies that it is a media company itself. Platform capitalism is a void, an all-seeing cipher: however large it grows, the organisation remains amorphous, evasive – 'agile', in business terms – and wholly without responsibility.

Which returns us to its founder. Zuckerberg embodies this same absent quality, a resplendent blankness paired with power too prodigious to define. As instigator of an 'open and connected' world, little is known about his private life. We know that he dropped out of Harvard, where he studied computer science and, interestingly, psychology. We know he has a wife, Priscilla, and two daughters named August and Max. We know he jogs regularly, because he posts photos of himself jogging on his Facebook page. We know that he worries about surveillance, because in a 2016 photo posted to celebrate the acquisition of Instagram, Zuckerberg's laptop camera is covered with a sticker.

We know that Zuckerberg owns a 700-acre estate on the

Hawaiian island of Kauai and has built a wall around it, one mile long and six feet high. We know he has filed lawsuits against hundreds of Hawaiians in order to settle disputes over land. We also know that Zuckerberg enjoys hunting his own meat, because he announced in 2011 that his new year's resolution was to eat only what he hunted. Pictures followed, soon after, of Zuckerberg grilling shanks of beef on a barbecue and holding a dead chicken by its legs. Zuckerberg, apparently, is so fond of meat that for his birthday his employees gifted him a meat-shaped cake. In 2012, he announced that his hunting challenge was over, via a 'connected steak' app called iGrill.

We also know – or, at least, I know, and believe that it will eventually be universally acknowledged – that Mark Zuckerberg is the progenitor of normcore. Normcore, the millennial fashion trend, is exactly what it sounds like: a celebration of mundanity in all its wearable forms. It involves ill-fitting denim, orthopaedic trainers, and oversized machine-worn cotton sweatshirts with ill-chosen fonts and dodgy graphics. It's the look of an ordinary, gormless, apparently decent person – someone so generic as to not show up on CCTV. First identified by trend-forecasting group K-Hole in their 2013 paper 'Youth Mode: A Report on Freedom', normcore celebrates clothing so cartoonishly basic as to resemble a character from *Seinfeld*, or from Microsoft's 1990s IRC client 'Comic Chat'. K-Hole writes that normcore 'finds liberation in being nothing special, and realises that adaptability leads to belonging'. Normcore functions as 'a path to a more peaceful life'.

Normcore has proven itself to be versatile, and unexpectedly enduring. It is at least in part a response to criticism of millennials as decadent, self-involved and lazy, because no one whose style is this utilitarian and bland could possibly be a bad person. Zuckerberg's style is normcore at its purest: he wears fleeces, hoodies, dad jeans and Adidas sandals, christened the 'fuck-you flip-flops' in the film *The Social Network*. His apparently unintentional personal brand slots into an archetype, that of the self-neglecting, socially inept Silicon Valley wunderkind. The idea is to dress like a slob, because you have no one to impress except other (invariably young, white, male) coders, who pride themselves on their ability to sit in darkened rooms for days on end without showering, typing code and drinking Monster Energy.

Mark Zuckerberg spent the better part of his first decade in the public eye wearing those same 'fuck-you flip-flops', and managed to get away with it without attracting too much scrutiny. His normality placed him beyond surveillance: he dresses in the same grey T-shirt and Facebook company hoodie every day, though in recent years the sandals have been replaced with trainers.

K-Hole describes normcore as 'youth mode', the infinite newness of a world where history has been flattened by capitalism and the internet; where everything has been made 'open and connected', and has already been seen. Facebook, as a service, flattens history, and reorders it on its own terms; the user is fixed in eternal distraction, present and past joined together on an ever-scrolling timeline which appears to stretch back beyond the year you first joined Facebook, following its users into the womb. Facebook's algorithm allows you to digest 'news' and 'updates' on a timeline, but not in chronological order; instead, Facebook calculates what you are most likely to want to see, and shows you this first.

'Fake news' breeds in these echo chambers, where stories too outlandish for the mainstream spring up again and again. They never fall down the Timeline, and they have no sell-by date, because they never occurred in the first place. Confused, we return to Facebook looking for answers, scrolling and scrolling without satisfaction. Facebook lulls us into dependence on the very service that keeps us in the dark.

With Facebook facing scrutiny for its influence over politics, truth-telling and human behaviour in general, the blankness of

Mark Zuckerberg becomes a model of millennial blamelessness. Like tech icons Steve Jobs and Bill Gates, Zuckerberg appears visionary, but not so idealistic as to hold political opinions anyone could disagree with. As a celebrity he tolerates a degree of surveillance, but his genius is that he is so fundamentally bland and uninteresting that we end up paying very little attention.

Who can fully comprehend how well Facebook knows us, and how little we know of Facebook? In my relatively short career as a tech writer, I have personally approached Facebook for comment on a number of occasions. I've even taken part in conference panels where a representative of Facebook was expected to appear. The reps did not attend, and Facebook HQ never replied to my emails. Facebook operates as a multinational Wonka factory: limitless information goes in and nothing comes out, apart from quarterly profit reports.

'To be able to blend, that's what realness is …'[2] I first learned about 'realness' from *Paris Is Burning*, Jennie Livingston's 1990 documentary film about ball culture and the LGBTQ community in New York. There, 'realness' meant replicating the everyday style of well-off, successful straight people in a nightlife environment, a survival strategy dressed as pageantry. Realness is complex in its motivations, and intensely political; it's about style, resourcefulness and chameleonic talent, but it's also a response to surveillance, and to the threats faced by marginalised people.

I also learned about realness from Facebook, where the term takes on a new meaning. Normcore is, in essence, a variety of 'realness' meant for straight people; an aesthetic response to technological surveillance, achieved through mundanity drag. Normcore touches a nerve in me, because personally I've never felt normal, for better or worse. I find myself intrigued and simultaneously repulsed by what Facebook's vision of realness represents.

Facebook convinced us that 'sharing' makes us better people, and fashion responded with normcore, a trend for innocuousness, wholesomeness and the appearance of virtue – albeit a distinctly neoliberal virtue, accessible mainly to the wealthy, and hinging on buying the right brand of fair trade coffee, tweeting support for the right causes, and 'sharing' – sharing everything, including your tastes, your relationships, your hopes and dreams and where you were last Saturday night. Normcore emerged in a world preoccupied with spiritual hygiene, where the diagnostic ritual of the confession booth has been transferred to social media platforms.

Facebook's advertising is peak normcore. Over the years they've published a steady stream of short video clips on YouTube, promoting new projects or recruiting new Facebook staff. This includes the iconic, unintentionally hilarious 2012 ad titled 'Chairs Are Like Facebook', later parodied on the show *Silicon Valley* as 'Sharing Is Tables'. Over one minute and thirty seconds, 'Chairs Are Like Facebook' presents the platform as essential to life worldwide, as essential as everyday things like doorbells, airplanes, bridges, the game of basketball, and, of course, chairs. The ad fails because of its misty-eyed, vaguely unhinged sentimentality, but at least it's transparent in its mission: Facebook wants to be as essential to your life as *chairs*.

Other Facebook ads take a more conventional approach. They're soundtracked by gently jangling stock guitar music, and take the form of montages depicting colourful, family-friendly scenes from everyday life. There is an overbearing wholesomeness to these images; it's as though Facebook has captured and branded precisely what it means to be agreeable, and happy. The unshared life, Facebook implies, is not worth living; here are blameless lives, lived in public, collected on the Timeline and shared with friends. One clip titled 'Show Your Stories' features a child ice-skating and an elderly man roller-skating, a graduation ceremony, a wedding and many, many shots of people of all ages dancing together. Another, titled 'Facebook Groups – Be Part of Something', shows folk music bands, preschool parenting groups, food drives, people flying kites, and a protest organised by Black Lives Matter. Yet another, titled '1,000,000,000 on Facebook', attempts to define what unites the site's users, and settles on the fact that all of them are human. A slightly smug female voice announces, 'That is a billion. And while it's a big number, today it somehow feels smaller than it ever has.'

I don't relate to Facebook's vision of life, or its idea of an 'open and connected' world. Perhaps I have too much to hide. Perhaps no one relates to it any more, although we have yet to produce an alternative. Every so often I try to quit Facebook, but I find the site so desultory and tedious that I can't even motivate myself to delete my profile. I go to the page made specifically for this purpose, but I always choose 'deactivate' instead, the temporary option. The 'deactivate' button was made for indecisive people like me; people who feel vaguely resentful towards Facebook, but who don't hate it enough to take proper action. What keeps me coming back? Most of my Facebook posts, pictures and even friendships there date from almost a decade ago. Why am I afraid of losing a self that I no longer relate to?

There is a sickness that comes from living under surveillance, one so subtle and gradual that it convinces the sufferer this is what health and happiness feel like. In her 2019 book on the subject, *The Age of Surveillance Capitalism*, Shoshana Zuboff compares it to the feeling of losing one's home: a dull, aching absence, and a sense of loss which Portuguese people call *saudade*. When I scan through my Facebook Timeline, or view one of those twee, occasionally horrifying 'Memories' posts from years before, I am confronted with loss rather than gain. I signed up to Facebook in college, convinced that if I didn't I would be missing out. But today I suspect that if I'd lived without Facebook, and without social media in general, I would have worked harder to make friends, and to build an adult identity, and both would be stronger and more lasting as a consequence. At the least I wouldn't have ended up in the conflicted state I'm in now, holding on to photos and 'Memories' from over a decade ago, resenting Facebook, and hating myself for my own ambivalence.

In a statement listing his priorities for the year 2020, Zuckerberg writes that 'In many ways, Facebook is a millennial company with the issues of this generation in mind.'[3] What are these issues? The need for attention and distraction? The need to compulsively 'share' one's life, and spy on one's friends? Through its core product, and later its acquisition of Instagram, Facebook made surveillance socially acceptable; it made being watched, by a corporation and by one's peers, seem aspirational and even morally good. As a writer and as a human I am challenged by the fact that my story has already been told, and archived, by Facebook, and Facebook has been profiting from the narrative of my life since long before I dreamed of writing it for myself.

My use of Facebook has affected me over the years, just like every other social media user, and more often than not its effects have been negative. Again and again it has prompted questions about the things I do not share. If I won't share them, *why* won't I share them? Are my inner thoughts somehow taboo? These questions, coupled with the triumphant vision of life my friends present on their Timelines, deepen the schism between my private and public selves. There must be something wrong with me, and one day that flaw will inevitably be made public, and will drive the world to shun me. I have in me something unwholesome, unlovable, un*shareable*; something hidden that there is no place for, because there are no hidden places any more.

For years, Facebook's monthly active users have far outstripped that of any other social network; as I write this, roughly one-third of the planet's population uses at least one Facebook product every month. Facebook has diversified enough to resist obsolescence; even if you have so far avoided Facebook itself, or deleted your account years ago, it's likely you'll still use Instagram, which has one billion monthly users, or WhatsApp, the world's most popular messaging platform, with two billion users.[4] As if to remind us of its own inevitability, in 2019 Facebook rebranded with an all-caps logo, 'FACEBOOK', and added itself to the names of its subsidiary apps ('Instagram from FACEBOOK', as one example, now appears on screen when the app is loaded). Facebook has even made its 'community' a physical reality, announcing plans to develop a social housing scheme in Silicon Valley. Willow Park, the largest development in the history of Menlo Park, where Facebook is based, comprises 1,500 homes, two parks, a 112,000 square foot hotel, a town square and 1.75 million square feet of office space. Facebook also pays a team from the Menlo Park Police Department, nicknamed the 'Facebook Unit', $2 million per year to patrol the area around its headquarters.

In his 2017 book *Move Fast and Break Things*, Jonathan Taplin identifies user 'interactions', based on figures from 2014, as work performed on behalf of Facebook HQ, and calculates an average 39,757 years of time spent on the network every day. He adds, 'That's almost 15 million years of free labour per year. Karl Marx would have been totally mystified.'

As leader of this unrivalled workforce, Mark Zuckerberg himself continues to resist interpretation. His power reaches into the improbable, far beyond the ambitions of the average tech CEO, or monarch, or dictator. Few critics name Zuckerberg as Facebook's body politic, a boy king guarding the networks and collective memory of billions, but this medieval implication is sustained by Zuckerberg's appearance and his behaviour. In interviews he is agreeable, equivocating, blank. Like the Queen of England, he never expresses his opinions in full, and in place of royal protocols he hides behind technology, which many in Silicon Valley would like us to believe is neutral and inevitable, with a will of its own.

Sometimes I fantasise about interviewing Mark Zuckerberg, not even to hold him to account, but to find out if his inner life matches his impassive exterior. I would ask him the questions Facebook began asking me from the moment I signed up. 'What films have you watched?' 'What makes you happy?' 'Add your relationship status.' 'Add a life event.'

Where does the wunderkind go, when he gets older? Does he settle into public life? Does he save the world? Or does he disappear entirely?

Rumours circulate that Mark Zuckerberg might one day run for US president. By touring America in 2017, and having photographers capture his visits with ordinary people, posted regularly to Zuckerberg's Facebook page, he already appears to have moved in this direction. His infrequent public appearances – delivering the keynote at F8, Facebook's official conference, and participating in 'fireside chats' at events including VivaTech and TechCrunch Disrupt – invariably leave the viewer cold, as Zuckerberg commits to little beyond his core message of openness and connection, which feels increasingly hollow with every iteration. The 'fireside chats', in particular, appear cynical; they're a way to venerate powerful, fundamentally unknowable leaders while appearing to humanise them. A fireside chat is very rarely revelatory, because those judged important enough to merit fireside chats are inherently protected from difficult questions. As an interviewing format it conjures mediated intimacy, yet it always takes place on a stage in front of thousands who have paid to be there.

What Zuckerberg represents, then, is a hive of connections, a cybernetic black hole that swallows up human behaviour and regurgitates it as ad revenue. He is a seer, a keeper of memories. He pressures us to share, and ostracises those who refuse.

He encourages us to watch each other, the way he watches us. In 2012 Facebook even instituted a 'snitch list', asking users to inform on friends who might be operating profiles under fake names (Salman Rushdie was targeted by the 'real-name' policy, along with a number of San Francisco-based drag queens who later took Facebook to court and won).

Zuckerberg made perhaps his most infamous public statement in 2010, in an interview conducted with David Kirkpatrick for his book *The Facebook Effect*, saying, 'Having two identities for yourself is an example of a lack of integrity.' What we know of Zuckerberg himself is hardly enough to constitute a single identity. We know him purely by his luxurious blandness, his posed photographs and his lengthy, ultimately meaningless statements about 'community building'. We also know, if we read Kate Losse's *The Boy Kings*, that Zuckerberg used to work in a glass office, in the name of radical transparency, but kept a secret private meeting room in the back.

Mark Zuckerberg has disparaged those who have two identities, yet he is used to having things both ways. Facebook refuses to be called a media company, yet the fate of every media company on earth is in some way tied to Facebook. Facebook declares no political affiliations, yet its decisions can swing elections, launch or suppress revolutions, and make or break political regimes around the world. Mark Zuckerberg donates to both the Republican and Democratic parties; in his home county of Westchester, New York, he is registered to vote but has indicated 'no preference'. He supports a more 'open and connected world', but also, in 2019, announced Facebook's 'pivot to privacy', shifting the focus away from sharing and the Timeline, and onto private communication through Messenger and WhatsApp (it's worth noting here that the co-founder of WhatsApp, Brian Acton, walked out three years after it was acquired by Facebook, leaving $850 million in unvested shares behind. He has publicly tweeted '#DeleteFacebook', and told the press he was coached by Facebook employees to mislead European regulators about what was happening to WhatsApp users' data).

In 2019, Mark Zuckerberg gave a speech at Georgetown University, making the case for Facebook as a defender of free speech.[5] Critics responded with scepticism; the speech felt more like an evasion, deferring responsibility for his product's problems and shifting their consequences back onto users. By framing Facebook simply as a platform for free expression, Zuckerberg puts off decisions about what is right and wrong – decisions that are left, in part, to moderators employed by Facebook around the world, who sift through the waves of disturbing material, the violence and conspiracies and sexual exploitation that proliferate on the site, and which spawn like the Hydra's heads.

What does any of this say about Facebook, and about its founder, Mark Zuckerberg? That he's used to eating his cake, and having it too. That he wants Facebook to be everything, and nothing, as long we continue using its products. That we will never know where Facebook stands, on anything other than its own limitless growth.

In *A Scanner Darkly*, Philip K. Dick's dystopian novel on paranoia, addiction and its effects, characters addicted to 'Substance D' undergo a change which causes their minds to split, leaving them with two sides operating independently of each other. After withdrawal at a 'New-Path' rehabilitation clinic, the novel's protagonist, Bob Arctor, is left cognitively impaired. Another character, Donna, says of him:

> 'He had no idea, and he hasn't any idea now, because now he hasn't any ideas. You know that as well as I do. And he will never again in his life, as long as he lives, have any ideas. Only reflexes. And this didn't happen accidentally; it was supposed to happen.'

I see something similar happening to us, as long-term Facebook users. First, there's the thrill of addiction: stimulation, distraction and maybe even new friendships, where users are brought together by a shared love of the same thing. After that comes paranoia, and the split: you don't know if what you're seeing is real any more, and whether, on Facebook, you can ever actually be yourself. You turn neurotic. You begin to perform and to self-fashion, checking in early and often and tending to this alternate self as your dependence on the platform grows. Finally, after prolonged use, you become a zombie: 'no ideas', nothing, caught in the filter bubble for life.

As Facebook's cultural cachet has declined, I've watched this process replicate itself in a different medium through Instagram. On Instagram, the visual takes precedence over text posts, or membership of groups; consequently, the bubble Instagram traps its users inside is an aesthetic one. For many, in particular female users, this has led to the rise of the 'Instagram face', a look created with make-up and, increasingly, surgical procedures which conforms to that of celebrities popular on the platform. For others, it's photos capturing wealth – cars, holidays, nights out and expensive clothing. In every case the same dynamic of power and envy kicks in; the person in the photograph will always be out there living, posing, enjoying a life worthy of Instagram, while you are alone with your phone, consuming their excitement second-hand.

It didn't always have to be this way; early social media was a canvas for self-expression, in all its strangeness and diversity. Avatars, pseudonyms and elaborately designed in-game characters were a default in online life, and 1990s forum culture, blog culture and even early 2000s social media all relied on users inventing names for themselves, with doxxing (revealing someone's real name, and their location) regarded as a transgressive and highly malicious act.

The false names adopted today by users of 4chan, 8chan and other communities of online ill repute were once a standard behaviour; in keeping with techno-utopian ideals, the internet was a place for self-reinvention, one where creativity was valued above truth. Online multiplayer games, blogs and anonymous forums encouraged users to experiment with identity, permitting forms of self-expression they couldn't find in offline life, including experimentation with style, power dynamics, sexuality and gender.

Facebook was not the first online platform to demand real identities of its users – in the late 1990s, as one example, Microsoft's 'Passport' product proposed a single, streamlined identity for users across a multitude of sites. But Facebook's impact in mainstreaming the use of real names, and stigmatising fake ones, cannot be understated; a certain aspect of online life was shut down, a version of the future cancelled. Facebook helped establish the confessional, earnest tone that came to define 2000s and 2010s social media: a sense of the internet as a confession box, where writing's worth depended on its apparently 'raw' and 'searing' veracity.

Mark Zuckerberg went on record against the idea of having multiple selves. The striking irony, here, is that at the same time he has gifted one-third of the world's population with a new online 'self', one that's rarely an accurate reflection of the person it represents in real life. Mark Zuckerberg has become a kind of latter-day Andy Warhol, every bit as blank and vastly more boring. We are his paltry Superstars, infinitely reproducible, but only on his terms. He has granted us far longer than fifteen minutes in the spotlight; Facebook owns the right to each user's name, likeness and image for life, for use in 'sponsored stories' advertising products to our friends.

Facebook, it seems, will never fully delete itself from this world; even if its user numbers fall and its profits decline, the company has acquired a diverse enough range of products, including Instagram, WhatsApp and Oculus VR, that it will endure long beyond its flagship platform. Similarly, you can delete Facebook from your life, but you can never fully delete yourself from Facebook; in many territories the 'right to be forgotten' doesn't apply, and even if you successfully remove your account, your anonymised log data will remain on its servers indefinitely.

I signed up to Facebook more than a decade ago; the time in which I've been a member is a significant stretch of my life. On Facebook I've crafted a self, just as I've crafted a self in writing, but that same identity never really belonged to me; it was dismantled even as it came into being, and sold off in parts to advertisers. This adversary, a somnambulist under the sway of a powerful magician, horrifies me with my own lack of 'self'-control.

Back in 2010, one night, I went to the Irish Film Institute to watch a film. I emerged from the cinema to find a party outside; the bar had been cordoned off with ropes, behind which were fifty or so smiling, oddly clean-cut-looking young people. They were, it turned out, employees of Facebook, on a company outing to watch *The Social Network*, David Fincher's film about Mark Zuckerberg and the genesis of Facebook itself. The film's negative portrayal of their leader didn't seem to matter; all publicity is good publicity when you get to Facebook's level of market dominance, and after the film's release its user numbers continued to grow.

A similar pattern has emerged with Facebook's political and legal challenges. When Mark Zuckerberg went before the EU Parliament, he listened patiently, laughed awkwardly, and avoided answering any questions in depth, before promising to 'follow up' with written replies in an email later. In Washington, appearing before the US Congress, his testimony lasted far longer, but still he managed to sidestep making any commitment to serious change in his organisation. He apologised for not having taken 'a broad enough view of our responsibilities' when it came to fake news and foreign interference in elections, but he also suggested that these problems could be fixed by investing further in technology.

Technology, it strikes me, is always technology's answer to the problems technology caused in the first place, much like suggesting the cure for an addiction is more of the substance you're addicted to. Since it was founded in 2004, Facebook has weathered political scandals, accusations of damaging its users' mental health, billion-dollar fines and confrontations with governments, and has always managed to do so on its own terms. Mark Zuckerberg has even given away his wealth on his own terms; he committed to give away 99 per cent of his Facebook shares over the course of his lifetime, but the donations will be distributed through a Limited Liability Corporation, the Chan Zuckerberg Initiative, rather than a non-profit. A non-profit would be subject to rules, and oversight, but Zuckerberg can do anything he wants with the shares, including electioneering and investment in any company he chooses.[6] Nor does the donation cut into Zuckerberg's wealth in any tangible way – he earns an average of $6 million per day.

It strikes me that Facebook is among the purest manifestations we have today of capitalist realism, defined by Mark Fisher in his 2009 book of the same name as 'the widespread sense that not only is capitalism the only viable political and economic system, but also that it is now impossible even to *imagine* a coherent alternative to it'. On repeated occasions in America and Australia, during outages of Facebook and Instagram, users have called the police demanding help. In Queensland, Australia, it's happened so often that the police force issued a statement asking the public to stop calling them.[7] Further proof lies in text which appears in my Facebook feed at least once every year with reliable frequency;

a 'takedown post' copied and pasted in the belief it will protect the user's digital rights:

> I do not give Facebook or any entities associated with Facebook permission to use my pictures, information, or posts, both past and future. By this statement I give notice to Facebook it is strictly forbidden to disclose, copy, distribute or take any other action against me based on this profile. The content of this profile is private and confidential information.

People look for protection from Facebook *through* Facebook, and the post is, of course, a hoax.

The takedown notice, itself a kind of bleak, unintentional meme, reveals that even users with a minimal grasp of digital privacy view Facebook with suspicion, and view their use of Facebook's products as an ongoing battle with an omniscient foe. It also reveals that they didn't read Facebook's Terms of Service before signing up (understandably – the document and its subsections add up to roughly 14,000 words), which state that when you upload anything to Facebook or its subsidiaries, you grant the company a 'non-exclusive, transferable, sub-licensable, royalty-free and worldwide licence to host, use, distribute, modify, run, copy, publicly perform or display, translate and create derivative works of your content'.

That this would happen in the first place is telling; the instinct is to post more content, rather than to quit the platform. What Facebook has is resilience, invulnerability, strength; the kind of all-consuming grasp on its customers that prevents them from even imagining life without their product.

Every year Facebook files hundreds of patents for technologies which might never reach the market, but which are nonetheless revealing of the company's plans. The number of patents filed rises every year; in 2012 it was 279, but by 2019 it climbed to 989.[8] Many of the patents are typical of a tech giant, relating to data processing and data transmission. There are also hardware innovations, including a drone that uses kites to stay in the air, and a series of functions for a VR haptic feedback glove. Others seem like predictable areas for Facebook to target: integration with Tinder, a peer-to-peer delivery app, an email system built into the Timeline and a Facebook-branded version of Chromecast, the smart TV device.

But others on the list branch into stranger things, including behavioural monitoring and prediction. Facebook has filed patents for software with the ability to judge users' personalities, to mine data from their phones to determine how many hours they sleep at night, to predict if you're in a romantic relationship with another Facebook user by tracking how many times you visit their page, to scan photos for appearances of brands and later sell this data to advertisers, to track people's eye movements, and to determine the camera that took a photo even if its metadata has been stripped. Most ominous of all is a patent for the ability to predict 'life events', including childbirth, college graduation, and, eventually, death. These products may not be available yet, but they indicate an ambition; Facebook wants to extend the reach of technological surveillance from the present day into the distant future, and to profit from every moment along the way.

Writing in the *London Review of Books* about Facebook and its use of 'retargeted ads', a marketing tactic which follows individual users through physical space using geolocation on their phones, John Lanchester concludes:

> What this means is that even more than it is in the advertising business, Facebook is in the surveillance business. Facebook, in fact, is the biggest surveillance-based enterprise in the history of

> mankind ... That's why the impulse to growth has been so fundamental to the company, which is in many respects more like a virus than it is like a business. Grow and multiply and monetise. Why? There is no why. Because.[9]

Neither Lanchester nor anyone writing for *The California Review of Images and Mark Zuckerberg* will go so far as to compare Mark Zuckerberg to a Lovecraftian monster, so it falls to me to argue the case. Zuckerberg represents the familiar – we know his face, and we deal with his product every day – but also the sublime, the unfathomable, the potentially horrifying.

Facebook's earnings come from our 'likes' and dislikes, our relationships and behaviours and connections. In this sense, Facebook is a machine trafficking in the things that make us human, and to understand its workings in full would mean seeing ourselves sold as humanity in bulk. Mark Zuckerberg's statements on free speech, made in public appearances, read as saccharine, inane and largely pointless, until you realise that it's in the interest of his business that everyone – the troll farms, the religious extremists and the neo-Nazis included – keeps using Facebook at all costs. Facebook is a platform where speech is free, because speech, whatever its content, will never fail to turn a profit for Facebook.

No one will be left behind, no experience left un-monetised. To understand in full the scope of Mark Zuckerberg's vision, his plan to mediate every moment of existence, would surely inspire bafflement, or horror in the average user, or perhaps a *Scanners*-style explosion of the head.

One cannot understand Mark Zuckerberg by seeing him. It's necessary, instead, to see what he sees. This brings to mind the God View enjoyed by many tech CEOs, the ability to see all your users, their actions, their current whereabouts and other data from behind a screen in your company's war room.

In 2010, Google's CEO Eric Schmidt described Google's God

View in a speech delivered at the Washington Ideas Forum: 'We don't need you to type at all. We know where you are, with your permission. We know where you've been, with your permission. We can more or less know what you're thinking about.'[10] This tension between permission – the terms and conditions we do not read – and its exploitation is key; we're assured that *we signed up for this*, and that surveillance will make our lives better. Precisely how that works, however, and how tech companies profit from surveillance, are kept from us as ineffable mysteries.

Uber, a more recent arrival in the pantheon of panoptic big tech, agreed to submit to two decades of Federal Trade Commission audits in 2017, after facing criticism for showing off their own God View of drivers and users on a giant screen at a party. Reportedly, Uber management were also using it to spy in realtime on politicians, celebrities and their own ex-partners. This was clearly an abuse of power, but the God View itself is a fact of life under technology; it's proof of the same ambient surveillance we've known about for years, ever since Edward Snowden went public about PRISM, but instead of state intelligence agents, here it was being abused by a crowd of drunk tech execs.

It's a challenge to imagine Facebook's God View in full. Certainly it includes our physical locations, along with our web of relationships, our place in the community Facebook claims to facilitate and build. It includes everything we opt into when we download one of Facebook's apps: the contents of our phone's address books and photo folders, our calendars and call histories, and our faces, with all their individual features, thanks to DeepFace, a facial recognition software owned by Facebook, and reportedly more powerful than that used by the FBI. It also includes a detailed record of our emotional states, because Facebook can gauge and even control its user's moods. This was proven with the notorious 'Emotional Contagion' study of 2014, in which Facebook's researchers altered users' moods by manipulating the contents of their news feeds, as well as a pitch to advertisers that leaked, offering the option to target users when they felt 'worthless', 'stressed', 'defeated' and 'insecure'.[11]

But a map of what Facebook knows about us might also branch into the metaphysical; it might include our hopes, our fears and uncertainties, and the secrets that play out between ourselves and our screens. Their map would stretch into the past and future, reaching both ends of the user's mortal timeline, and beyond.

We may never know how much Facebook sees of us, but the site placates us with a miniature God View of our own, every time we log in. Facebook's user interface is an overview of our lives: along the left-hand side are our groups and events, the centre is a news feed, and the right-hand column is a buffet of our friendships. Facebook imparts its dehumanisation of its users *to* its users in turn, encouraging us to view our social life as a game of alerts and dialogue boxes. Facebook breeds solipsism in the user. Everything is arranged for you to 'like' and acknowledge and comment on. Everything is happening to you. You can pick up conversations with people and drop them again in seconds. You will never see yourself on a list, as part of someone else's menu. You will always be the centre of your world.

Bearing in mind his God View over 2.07 billion people, perhaps it is useful to think of Mark Zuckerberg himself as a kind of god. Perhaps he's one of the Great Old Ones: amorphous, malevolent, hungry, capable of inducing madness in anyone who sees his true form. Or perhaps he's more like the god of Genesis: he has given us a garden of our own, as long as we keep to his rules.

Certainly there are those who already treat him as a god; each of Zuckerberg's Facebook pictures attracts thousands of adoring comments. The team of twelve employees who manage his public profile have likely filtered out all the negative ones; those that remain are so adoring and supplicant and effusive that they make for uncomfortable reading. In one picture, taken during his much-publicised 2017 tour of America, Zuckerberg eats pie à la mode alongside truckers at a diner in Iowa. A comment, one of 310 posted below, reads 'Mark Zuckerberg you are such a down to earth & amazing person. That explains why God blessed you with so many things. Please stop by at North Platte, Nebraska. Take care.'

To 'see' Facebook, one must register for an account, and eventually be owned in some small way by the service. For every image posted, or comment made on Facebook, the service claims you for itself. This kind of surveillance can't be negotiated in pieces; you're either in completely, or you're out.

What will happen when Facebook runs out of new users to convert? What will happen when Facebook's filter bubble begins to work against it, when – accustomed to copying each other and 'liking' the same things as the rest – we stop demonstrating the unique behaviours and interests that once made us useful as ad data? What about when we have nothing to offer, when we – like Zuckerberg – become so bland, so beaten down by peer pressure, as to be 'normcore' completely, and beyond surveillance?

Is Zuckerberg himself an augury of this future? Is he a benevolent god, a genius, or a warning? Is he Moby Dick, a sprawling white void of American power ('Is it that by its indefiniteness', Melville asks, 'it shadows forth the heartless voids and immensities of the universe, and thus stabs us from behind with the thought of annihilation ... a dumb blankness, full of meaning, in a wide landscape of snows?'). Might it be that Zuckerberg is not white, but 'cosmic latte', the distilled hue of the universe?

It's painful to think of an ordinary young man having so much power, and money, and information. Easier to think of him as a robot, a savant, or a meat-hunting superhero. Easier to conceive of Mark Zuckerberg as a machine.

A machine, or a cyborg, at least; one entirely comfortable with technology, and the change it brings to his life. This is the metaphor that fits; this is why he thrives; this is the difference between Mark Zuckerberg and me. In Facebook, Mark Zuckerberg offers us the chance to do what he did long ago: to fully commit to the network, and to a future as something other than ourselves.

To build a relationship with someone online is to draw them into a web of self-fashioning, and to consent to mutual surveillance. With Facebook, the same thing has happened: Facebook has seduced us, and encouraged us to rebuild ourselves. As their core product wanes, and messaging apps take precedence over the Timeline, Facebook will still be there, ushering us through the online experience, pre-empting our needs and meeting them at every stage of our relationship with its products. We've been welcomed into a placid blue-and-white world where someone is silently watching, organising our lives, making us feel like we deserve their attention. It is as though Mark Zuckerberg loves us, more than anyone else ever will.

Perhaps this is why we tolerate the figure of Mark Zuckerberg. He will always be more boring than we are. He makes us feel like we are worthy of surveillance.

We accept the friendship of MySpace Tom, the all-seeing eyes of Zuckerberg, and the promise that Elon Musk will someday find a home for us on Mars. We sell our lives to Google, because in return they make life easier by breaking it into actions and data segments. There are those among us who bask in surveillance, who dream of an NSA agent out there watching, checking in on us from time to time.

A passage in *The Boy Kings* comes to mind here, written not about Zuckerberg himself, but about a video chat conversation between the author and someone she cared about. On screen, the feelings are simpler, easier to access, and more readily extreme.

It becomes difficult to separate human from machine, mediation from emotion, selfishness from love:

> I think we could tell him we love him because he was so far away, and to love him is to love the technology that allows us to speak to him anyway, safely, intimately, from afar. Our technology, ourselves: For us, at the heart of this revolution, they were increasingly the same.

Pink Light: Notes on Six Vaporwave Albums

> If I asked her why she was always trying to need more than she needed, she'd say that borrowing brought you closer to other people, while buying mostly made you lonelier.
>
> *You Too Can Have a Body Like Mine*, Alexandra Kleeman[1]

1. Macintosh Plus, *Floral Shoppe*

Vaporwave is happening to you. It's a way of seeing, of being, pitched somewhere between the internet and real life. Vaporwave is an aesthetic, a set of politics, an aggregated history consumed through a screen. But before that, vaporwave is a music genre, which first appeared in the early 2010s.

Vaporwave found me at a time when my musical tastes had narrowed, through a mixture of habit and apathy. I needed something to listen to while writing late at night. Ideally it would be music without words, which I found distracting; soft, long-playing tracks without sudden shocks to the eardrums.

I had cycled through the same Aphex Twin albums on YouTube for years, feeling guilty each time I returned for not being more adventurous. Then one night I clicked an unfamiliar link in the sidebar, and this led me to a trove of lo-fi and 'vapor-trap' mixes posted by accounts with names like 'immortalyear', 'E m o t i o n a l T o k y o' and 'STEEZYASFUCK'. This was my introduction to vaporwave. The videos were endless, all of them roughly an hour in length and containing samples from the 1980s and 1990s, old J-pop records and elevator music. Soon after this, the algorithms led me to the pink-panelled floor and the bust of Helios, the vaguely forbidding visuals which feature on the cover of vaporwave's best-known work.

Floral Shoppe was vaporwave's breakout hit, published on the website Bandcamp in 2011 by an American producer called Macintosh Plus (this artist is also known as Vektroid, New Dreams Ltd and PrismCorp Virtual Enterprises). Unlike other vaporwave albums, *Floral Shoppe* is demanding; I've never been able to just leave it playing in a tab in the background. From the start there's an insistent saxophone, and a man's voice, glitching and crooning, 'I love you so ...' His love has become a system error; it sounds like bad memories, or regret. The second track, 'Lisa Frank 420/ Modern Computing', is detached and creepy, like the theme song from an out-of-date children's show. A chugging, locomotive beat backgrounds a sample of Diana Ross singing 'It's Your Move'. Her voice is dramatically lowered, her words blurred and woozy, and the track itself, with its lyrics about unrequited love, becomes a hymn to passivity.

It makes sense that the track vaporwave is best known for would be about longing, and about watching from a distance. Vaporwave is native to the internet, where nothing is ever real enough to grant full satisfaction. The genre documents technology's reach, and its manipulation of history; it delves into the past and restores it as warped, with the human emotion filtered out. The lyrics frame love as retail: Ross complains that the person she's singing to isn't 'buying' what she has to offer. The track sounds like it's being played on broken machinery, like we're travelling through a tunnel, or through the wires of the internet itself. Something is taking place, somewhere, but we're too exhausted to care.

Other tracks have a rhythm, a sense of motivation and purpose. 'flower specialty store' and 'library' conjure inorganic landscapes, a digital realm made audible. 'ECCO and chill diving' really does sound like liquid, crystalline and sad, while 'Geography' and 'Mathematics' are like rumblings from a video game's hidden levels. 'Wait', the final track, unexpectedly recalls nature; birds sing in the background, while a keyboard rings out, shrill and ecstatic, apparently waking us from a dream.

Vaporwave takes its name from vaporware – tech products announced to the public, often with great fanfare, before failing to be released and disappearing entirely. Throughout the 1980s and 1990s the term was associated with Microsoft, and the long-delayed, underwhelming video game *Duke Nukem Forever*. In this way, vaporwave has its roots in disappointment, and scepticism of technological hype.

Floral Shoppe, and vaporwave in general, is also closely associated with the term 'A E S T H E T I C'. This describes a visual style popularised on Tumblr, drawing on late twentieth-century pop culture, tech graphics and other corporate debris to assemble a glazed, quasi-science fiction look. A E S T H E T I C is eclectic yet oddly rigid in its parameters; threads appear frequently on Reddit debating exactly what A E S T H E T I C is, and rarely reach any definitive conclusion. It is both lurid and cute, encompassing anime, old ads for products like Coke and Pepsi, ancient statues and neon lights (pink – almost always pink, coupled with shades of indigo blue and turquoise). A little louche, a little cyberpunk-meets-*Miami Vice*, A E S T H E T I C summons a mood of nostalgic exhaustion, one which complements vaporwave music perfectly.

Multiple vaporwave subgenres have emerged, including hardvapour, late-nite lo-fi, signalwave, mallsoft, casinowave and Simpsonwave, all of which rely heavily on the things from which they take their names. Casinowave replicates the sounds and experiences of a casino. Mallsoft does the same thing but with malls. Simpsonwave incorporates clips and visuals from *The Simpsons*, itself a cultural archive for 1990s children, while signalwave, also called Broken Transmission, produces an effect like channel-hopping by weaving in clips of random audio. Vaporwave's references can be corporate or commercial; they include malls, advertising campaigns, 'business casual' fashion, and brand names like 'Crystal Pepsi' and the fictitious 'Doldrum Corp.' Artists change their names frequently, an old-web gesture, which makes vaporwave seem like a tulpa, an entity manifested from thoughts, and born of an anonymised cultural mood rather than from an individual person.

How much of life is algorithmic now? I keep coming back to vaporwave, perhaps because YouTube knows that I enjoy it and recommends more. The time I spend listening to it is also time I spend alone with my computer; vaporwave complements the night, the retreat into memory, and the brain's eventual sorting of events in sleep – rapid, then slow, then rapid again. YouTube, where more music is played globally than on Spotify, Apple Music and every other streaming service combined,[2] is a hive of vaporwave music, enough that, with the help of the autoplay button, and enough battery life, vaporwave could soundtrack the rest of my existence.

It took me a long time to share with other people that I listen to vaporwave, not out of embarrassment but because of the ritual of the thing; it never appeared in my life outside my computer. It took even longer for me to think of myself as a vaporwave fan, despite playing it for years, always alone, and almost always when I had a deadline to meet. It's jarring, somehow, to consider listening to it outdoors; these are soundtracks to loneliness, late-night companions that luxuriate in the psychic ties between the individual listener and the internet. To me, *Floral Shoppe* summons feelings of numbness, futility, exhaustion, but peacefulness too. It speaks to my free-floating passivity, an acceptance of a world I cannot touch, curated for me by machines.

2. Blank Banshee, 0

One year after *Floral Shoppe*, another key vaporwave album, Blank Banshee's 0, appeared online. The cover features a hollow, digitally rendered face, almost certainly that of *Tomb Raider*'s Lara Croft, placid and glaring despite having only empty holes where her eyes should be.

The opening track, 'B:/Start Up', samples start-up sounds for both the Apple Mac and Windows 95, the latter a 3.25-second clip originally composed by Brian Eno. Comments below the video mention being stoned, although personally I associate this album more with ecstasy, and the bizarre synaesthesia-type effects it can sometimes bring on – this happens with 'Wavestep', in particular, which I love more than any other vaporwave track. There's an optimism to this album, a distinctly synthetic beauty, paired with eerie lyrics about a 'bed of glass' – broken glass, we presume – which cuts the sleeper.

Later tracks have ominous names, like 'Hyper Object' and 'Bathsalts', which can be best described as a downward spiral, or a flock of metallic birds attacking your eardrums. But 'Cyber Zodiac' sounds clean, like the taste of chewing gum, or the smooth peripheries of objects. There's a sequence of notes at the end of this track that gives me so much hope it is almost painful, a sound like dawn in an open-world video game. The final track is called 'B:/Shut Down/Depression'. It sounds like the calmest place on earth.

Identity, in vaporwave, is a malleable thing; unlike the rest of the internet, where real-name policies have dominated the last ten years of social media use, the genre favours dissimulation, reinvention and irony. The aliases serve as a throwback to the online past, and give creators the freedom to be sad, sinister or absurd in ways we rarely see on Instagram, Facebook or Twitter. They also take ego out of the equation; vaporwave is a rare modern example of a pseudonymous online community that gets along.

The public face of Blank Banshee is a masked figure wearing a hooded jacket. The mask has empty spaces for eyes and an eerie grin, covered in pieces of mirror like a gurning disco ball. Sometimes Blank Banshee also appears wearing a balaclava, like a member of some new and disaffected deep web iteration of the IRA.

I love that about vaporwave; its sense of deadened camp, which the names of its artists, albums and tracks speak to: Far Side Virtual, Saint Pepsi, Megathrust, *Googleplex Bionetwork*, Laserdisc Visions, Eyeliner, Luxury Elite and *Computer Death*. You need to go down a wormhole just to find a lot of the tracks; perhaps this is why so many vaporwave albums reference being lost or suspended in specific locations (*Palm Mall*, *Purgatory* or *End of World Rave*). They seem almost to map the internet, its niches and topographies. Possibly my favourite album title is *WEBINAR*, which takes a corporate neologism and turns it into something new. I often receive invitations to webinars; they go straight to my spam folder.

On 20 February 1974, Philip K. Dick was at home recovering from dental surgery when a woman with dark hair and a golden, fish-shaped necklace arrived at his door to deliver painkillers. Dick watched as a beam of pink light burst from her pendant, triggering visions that endured in the months that followed. Pink light appeared, again and again; the colour of clairvoyance, he believed, and divine intelligence.[3]

Pink light is central to vaporwave, and is linked, I believe, to its core sense of apathy. Pink – 'millennial pink' – is the colour most often associated with my generation, the generation to which a large number of vaporwave producers and fans belong. It's likely this mainstream pink trend was inspired, at least in part, by vaporwave; the colour appeared on the cover of *Floral Shoppe* back in 2011, and in 2016 Rose Quartz was named Pantone's 'Colour of the Year'.

I write this under a pink light in Peckham, in London, in a bar on top of a car park while waiting for friends to arrive. I'm visiting for a few days, and had to climb four flights of stairs to get here, and those stairs, and the halls around them, were all painted a specific shade of pink. It's called Baker-Miller pink, vivid and Barbie-esque, the pink of plastic toys and nail polish. It's a shade that appears frequently in vaporwave imagery, like the covers of *Floral Shoppe* and Blank Banshee's *Mega*.

This colour is controversial; it's been used on the walls of prisons, holding cells, psychiatric hospitals and schools. It's said to calm and even physically weaken the beholder. A site called 'Exploreyourspirit.com' tells me that, in New Age colour theory, 'pink in its clearest form in the aura represents joy, romantic love, and an optimistic view of the world', but that 'a washed out and faded pink or a muddy pink represents a person who has not developed fully from their childhood ... They often struggle with codependency and feelings of helplessness.'

Like pink, like the mood, sound and aesthetic of Blank Banshee, I often detect in myself and my peers a sense of futility and numbness, as though the same thing that gave us the world – the internet – has also rendered us powerless. Vaporwave gives voice to this state, and we retreat further into its smothering cocoon, distracted, defensive and vulnerable. 'Lay me down in a bed of glass / Cut myself as I turn.'

3. 식료품 **Groceries, 슈퍼마켓 *Yes! We're Open***

It's rare to find a concept album, let alone a well-regarded one, themed around the layout of a supermarket, but that's what 슈퍼마켓 *Yes! We're Open* is. Over thirty-four minutes, it replicates the experience of food shopping, with tracks named after numbered aisles, ending with 'Checkout (Have a Nice Day)' and 'Bronze-Level Store Loyalty Card'.

슈퍼마켓 *Yes! We're Open* draws heavily on muzak, tracks made to pulse ignored in the background of retail environments, steering visitors along a guided tour towards the cash register. I've seen this album labelled bleak, and, in other reviews, as an escape into childhood fantasy, but personally I think of it as neither. It feels apprehensive in parts and in others blandly dynamic, summoning emotions only to build, ultimately, to nothing. Mundane retail sounds are rendered strange; we hear panpipes, salsa, a droning saxophone chorus, and the background hiss of a supermarket Tannoy. 'As always,' a voice says, 'we thank you for shopping with us. We are open twenty-four hours a day for your convenience.'

'Interlude (Lost in the Freezer Section)' is a glacial arrangement of sound effects and mournful saxophone. A sample is taken from The System's 'Don't Disturb This Groove', a number-one hit from 1987, the same year the pop star Tiffany toured malls around America. Some of the tracks on 슈퍼마켓 *Yes! We're Open* sound like they were played and re-recorded in an empty room, and I get the sense that, however good my headphones, and however loud I play it, this album will always retain a certain soft-focus quality.

This album romanticises its subject, implying – through its hazy tone, and its very existence – that this mundane retail experience is already a thing of the past. Perhaps grocery shops, like all things, will be 'disrupted'; three years after 슈퍼마켓 *Yes! We're Open* was posted online, Amazon launched AmazonFresh Pickup, a drive-by grocery service available for same-day or next-day collection, with plans to gradually roll out the service worldwide.

슈퍼마켓 *Yes! We're Open* is categorised as mallsoft, the vaporwave subgenre which evokes the sounds and atmosphere of shopping malls, and which obliquely references their current, precarious state. Technology is treated with ambivalence in vaporwave; it enables music, and community, but it also accelerates time and loss. Mallsoft confronts this conflict head-on; it's a dreamy, idealised online evocation of a thing that's slowly being killed by the internet.

Over the last decade malls have declined, their sales and visitor numbers falling. The preference for shopping online has sparked a rise in 'showrooming', the practice of examining items in person but buying them later online. Bricks-and-mortar shops operate as museums of things shoppers will buy later, with a trend for 'experiential' retail – shops filled with cafes, beauty salons and other services designed to lure people through the door.[4] In the mall, the real world becomes like a video game; algorithms are used to determine tenancy rates, to project sales figures and to track footfall, as well as determining the location and design of malls themselves, which continue to be built despite diminishing returns.

If online shopping is killing the mall, then vaporwave is already in mourning. The cloying optimism of albums like 슈퍼마켓 *Yes! We're Open* or *Palm Mall* by 猫 シ Corp. is only semi-satirical; while it assumes a certain level of anti-consumerist distrust in the listener, it also assumes a shared sense of nostalgia for the mall experience.

I worked at a mall throughout 2010 – a long time ago, but even then we could sense that it was declining. There was no muzak playing in the background; instead, we had a medley of almost-contemporary pop tracks from the year before, or the year before that, which played in rotation from 9 a.m. until 9 p.m. At the Dundrum Town Centre, a jagged, 140,000m^2 edifice in the south Dublin suburbs, overwrought pop hits like Leona Lewis's 'Bleeding Love', and 'Love the Way You Lie', by Rihanna and Eminem, droned throughout the four-floor department store, haunting and occasionally tormenting me as I worked on commission at a beauty counter. I lacked all qualifications as a make-up artist, but I discovered that I was good at persuading women to spend money. I 'traffic stopped', pulling customers over to the counter. For a while I even genuinely enjoyed it; I was calmed by the order and routine, the sales patter and the glassy, antiseptic look of the place.

This was shortly before blogs, and, later, YouTubers restructured the beauty industry completely. Our priority was getting the product on a potential customer's skin; tactile experience was something the internet couldn't provide, however carefully the YouTuber swatched products on their hand and waited for the camera to focus. Tactility was a source of pleasure in my work; it grounded me, helping to stave off depersonalisation. I liked polishing the counter and lining up the lipsticks in their holders, labelled with names like 'Autumn Glow', 'Desert Rose' and 'LA Diva', and cleaning the brushes at night, running them under the tap one by one and watching cloudy, skin-coloured water whirl into the drain. During those months I repeated this action so frequently that I began to dream about it, about laying out the brushes on kitchen paper to dry, and about the slack texture of skin around older women's jawlines, the way it vibrated, slightly, as I applied mineral powder foundation.

Vaporwave is happening to you, and back then retail was happening to you too; mallsoft replicates the soothing mindlessness of that same customer experience. When I sat women in the chair and used the €50 kabuki brush on their faces, I saw them slip under. There was even a technique our rivals practised, at Benefit, the cosmetics brand across the hall, called the 'Benebubble'. They sprayed the air around you with perfume, then demoed six products on half your face before asking you to try the other half yourself. The odds were, apparently, that people would buy at least two products every time.

Back then I was a human algorithm, persuading women in place of a screen, and picking up on behavioural cues the way Amazon and Google Shopping do now. We were given company handbooks and onboarding sessions, and later I was shown a curious hour-long slideshow about self-tan sales, and the phenomenon of 'Tanning Thursday', during which women retreat to their bedrooms to spray themselves and wait for their self-tan to dry in preparation for the weekend. We were trained to be attentive and assertive, adding up figures, calling a €40 face powder a bargain. I've noticed that beauty counter staff are nowhere near as pushy today – perhaps the companies they work for have decided there's no point.

I've read about mall walkers, retirees in America who use local malls for safe indoor exercise. Originally the malls didn't know what to do; they agreed to open early to accommodate them, and tried to ply them with coupons to make them spend money. Sometimes – frequently, in fact – the mall I worked in was empty apart from the odd elderly person out for a stroll. My job was to seize on women the moment they appeared; if I worked the late shift, or early in the morning, I'd risk going hours without making a sale. I found myself wondering about the economics of the place; how did it make sense for shops to be open, when many of them only made sales on the weekends? I'd stand slouched at the counter for hours, and occasionally women would appear, alone or in pairs, almost always exhausted, and confused-looking, staggering towards me. Invariably they had no interest in make-up; they only ever asked for directions to the exit.

I remember this as a time of indecision, and gentle monotony. My first 'proper' relationship, one that was volatile and messy, had finally ended, and I knew that I wanted to get out of Dublin soon. In the meantime I was earning money at least, saving it for some point in the near future when I'd have a plan.

At night, sometimes, I wondered if my work was morally

questionable. I told myself that customers knew what they were signing up for; that they didn't come to the mall, or the beauty counter, unless they were willing to buy things. But that didn't seem entirely true. Could a place – a looming, cavernous place like the mall, haunted by softly droning music – make the rules, and justify me essentially preying on people? At the end of each day I'd walk through the staff exit and feel like I was casting off a persona, returning to my real self. Eventually the job became too much for me, and I quit.

Lately I often read of a 'retail apocalypse', one I am likely helping to speed on by rarely, if ever, buying in malls or bricks-and-mortar stores. I prefer charity shops, and shopping online – in high-street shops I get overwhelmed by the inability to find precisely what I'm looking for. Malls are a novelty; in everyday life I very rarely go near malls any more, yet while writing this piece I'm inspired to visit the biggest mall in Europe.

I'm already in London covering a tech conference, and so on the last day I take the Central Line to Westfield, the vast shopping mall in Shepherd's Bush. It cost £1.6 billion to build, and its retail space alone covers as much land as thirty football pitches.

Once I'm inside the entrance vanishes discreetly, ensuring I can't find the way out again. I pass luxury brands – Prada, Burberry, Gucci and Louis Vuitton – before turning onto a dizzying concourse, with shiny white floors and a vast canopy roof made from triangular glass panels. Screens hang from the ceiling advertising valet parking, and I notice a stand for charging phones, and a water fountain with a sign reading 'REHYDRATE HERE'. I move between soundtracks; in one hall it's nondescript jazz, in another it's George Ezra's 'Hold My Girl'. Around a corner this gives way to a chirpy, high-pitched voice that might once have been a woman's, but which has been distorted to sound like it comes from outer space.

I pass a shop called 'Urban Revivo', and one selling 'EMS Pulse

Core Technology', wearable devices designed to build muscles while the wearer sits still. I am lost in the mall already, even though I can't have arrived more than ten minutes ago – blame the Gruen transfer, the effect caused by passing from outdoors into a mall environment, which creates instant confusion in shoppers.

When I worked in retail, having to put my phone away for eight hours per day was a special kind of agony. Now I hold my phone out in front of me, filming my own vaporwave mall tour. I notice that everyone around me is holding up their phone too, taking selfies, or filming, or in video conversation with someone else. There's a procedurally generated look to this place, all glass and white plastic and gleaming geometries.

An intercom announces a seventy-two-second silence for those who died at Grenfell Tower. The music stops, leaving only footsteps and the ambient hum of the escalators. I continue, in silence, past schoolchildren walking in slow-moving packs, past shops selling featureless clothes, and featureless homewares, and past a stall selling 'Endura Roses'– flowers designed to last for ever in glass cases.

The mall feels empty, but then, by design it's likely too big to ever appear convincingly full. I can't imagine finding anything I'd need here, or, for that matter, anything I'd want. Words make me dazed; 'Extreme Hazelnut', 'Full Body Wax', 'Massage Angels', their stand covered in further phrases arranged in a decorative word-cloud; 'Energy', 'Happiness', 'De-Stress', 'Unwind', 'Positivity'. The halls meander, and my feet start to drag. Finally, I find another exit and walk to the station, and as the train pulls away I notice I'm feeling relaxed for the first time in hours.

What did I take from my visit? That shopping malls are like rifts in space–time, portals to a time and place where they're not yet obsolete. In the mall, the sense of escape no longer felt pleasurable; it felt like oblivion, like delusion. Still, there's a novelty to these places, in the smooth surfaces, like the inside of a spaceship. Reality takes on a bubble-wrap quality, a minty freshness like a ketamine trip. Vaporwave conjures that feeling through a screen, in tracks playing over old 'mall tour' videos. It reminds us that the mall is dead, and yet now the mall is everywhere.

Do vaporwave musicians earn money for their work? It's doubtful. If they gain any degree of success they attract copyright issues on YouTube, due to sampling, and are made to take their music down. In this sense, vaporwave stands at odds with commercial culture, even if it fetishises it too.

Now I am haunted by Instagram posts, advertising things I saw at the mall but decided not to buy. There are no goodbyes at the mall, only introductions. If they show something to me enough times I'll probably comply; I'll be lured in, and eventually I'll purchase. They'll get me at a time when I'm algorithmically vulnerable, knowing how lonely I am late at night when I use my computer.

4. 猫 シ Corp., *News at 11*

Certain drugs create a sense of oncoming revelation. Vaporwave can do that too.

News at 11 is a concept album, not about malls, or grocery shops or casinos, but about a world event. The first side of the tape has nine tracks, and the second side has eleven (vaporwave albums are frequently sold on old-fashioned cassette tapes). If this wasn't enough of a hint, the cover image shows an American flag, warped and seen through an airplane window. It's a portrait of a nation on a precipice; tracks contain samples of newscasters, daytime TV hosts and weather presenters on the day the Twin Towers fell, captured moments before tragedy struck.

'I don't know, it's kind of quiet around the country,' says a voice, backgrounded by lazy snare drums. 'We like quiet. It's quiet. Too quiet.' The saxophone starts to sound like a death march, militaristic and laboured. 'It's beautiful outside, perfect September day and lots of sunshine ... I'm going outside today.' Then come French horns, and an odd, jammy piano like a bad love song, oppressive, like cigarette smoke breathed down your neck. A voice from the morning show *Daybreak* discusses the stock market value of Nokia and Motorola, instantly recalling early experiences with technology, for some listeners at least. Samples come from The Weather Channel, Good Morning America, Downtown, Channel 4, Financial News and Evening Traffic, because *News at 11* is as much about media as it is about national tragedy. Then a voice starts talking about someone living in neurotic isolation; it's an interview with the author of a book about Howard Hughes, who describes him, for all his flaws, as 'the most amazing man who ever lived'. 'I have got to interrupt you now,' the host says. We don't hear what comes next, but we know.

猫 シ Corp. was born in 1989, the same year as me, and his music conveys all the frustrated romanticism and futility of a child of the 1990s, or, more specifically, the very end of the 1980s.[5] I remember being in school when I first heard about 9/11, going home later in the day and watching it on TV, and feeling – just like everyone says, and writes – like I was watching a disaster film. *News at 11* commemorates the puncturing of the American dreamworld, the media shepherding a long, soporific slouch towards a forever war. Like Ottessa Moshfegh's *My Year of Rest and Relaxation*, in which the protagonist doses herself with antipsychotics and sleeping pills for a year in the run-up to 9/11, it pinpoints a moment of numbness, self-induced, and a media culture that existed not to inform viewers of the latest world events, but to seal them into stasis.

Judging by the comments, listeners return to *News at 11* annually on the day itself, ever since it was released in the Trump-year of 2016. It's likely they're familiar with the news clips sampled on this album from compilation videos on YouTube. The album itself serves as a compilation, an archive, a tribute and a work of theory in one. It interrogates vaporwave, asking if the genre can only truly be appreciated by those who have lost something. This is the sound of ghosts, not of those who died, but of the culture that died with them. 'Album belongs in a museum', one comment reads, but this album *is* a museum, recording a world that might never actually have existed.

News at 11 is a memory-processing ritual, like the defragmentation of a hard drive. It's a lament for the end of context, because none of these clips would mean anything were it not for what came next. Who runs the Timeline? Who gets to tell the definitive version of history's events? 'The extended forecast. The extended forecast ...' We're near the end now, but the weather's not over yet.

How Americanised I am, how reliant on platforms owned and run by American men. How the internet lulls me into an inherited stupor, the way television did for past generations. There's a numbness I live with, that sometimes drives me to self-sabotage and self-harm. There's a sense of speeding towards a future point, a warning on the verge of being expressed, but looped back and glitching, lost in the cybernetic dross.

In *Capitalist Realism* Mark Fisher writes, 'Capitalism is what is left when beliefs have collapsed at the level of ritual or symbolic elaboration, and all that is left is the consumer-spectator, trudging through the ruins and relics.' He describes history being flattened, decontextualised and defanged; 'In the conversion of practices and rituals into merely aesthetic objects, the beliefs of previous cultures are objectively ironised, transformed into *artifacts*.'

A while ago, Fisher became my favourite interview subject: he spoke to me for a piece about England as a 'boring dystopia'. His writing lit up the world for me; it confirmed to me that life was as strange and inhumane as I suspected it was, but also that everything could be unravelled, and exorcised, and that depression could be used as a critical tool. The night I found out about his death I came home early from a party. It was a few days later; I had been avoiding social media, and so hadn't learned the news when it first became public. I sat in the dark and put on an hour-long vaporwave mix with a looped anime video. It showed a girl alone at night with rain outside, sitting on a bed, face buried in her hands and crying, just like me.[6]

5. Chuck Person, *Eccojams Vol. 1*

In 1958, when she was twenty-six years old, Sylvia Plath wrote the 'The Times Are Tidy', a poem about the oppressively cosy tone of life in post-war America. It describes a period when conformism, consumerism and anti-communist sentiment all helped to define the public mood. Yet Plath's lines always remind me of life with technology today; the hypnotic loop of a stuck record, and, later in the poem, the rotisserie that turns by itself conjure the predictability of algorithms. It's a poem about stasis, apathy, possibility sacrificed in the name of convenience. It takes place in the aftermath of innovation: the point when technology stops being exciting and starts to monopolise its user's life.

In vaporwave, the sound of a stuck record – and a listener too depressed to fix it – becomes a distinct musical feature. YouTube comment sections still debate whether *Eccojams Vol. 1*, released in 2010, invented the vaporwave genre, or whether it created its own genre which runs alongside vaporwave in tandem, called 'eccojams', characterised by looped samples and echo effects. Definitions are vague, here, and mutable; critics project wildly varying opinions onto *Eccojams*, and vaporwave in general, and often end up rendering a DIY art form completely inaccessible. I have always struggled with music criticism – years ago, in my early twenties, I tried to write music reviews for a magazine and promptly gave up – and part of me loves how vaporwave calls its bluff; it's a musical genre which demands study, and interpretation, and which turns out to possibly not be music at all.

With *Eccojams*, its affective powers depend on the context in which it's consumed. It depends on the mood of the listener, the time of night, one's solitude and proximity to machines. This album demands a lonely place – you have to be lonely in order to appreciate it – where no one is around to judge you for listening to something so strange.

Eccojams is named after the 1992 game *Ecco the Dolphin*. The cover art is a decoupage; the name 'Ecco' is printed down its sides, like a logo on a tracksuit, with 'MEGA' printed in a Sega-esque font, an illustration of a shark, and a blurred newspaper page reading 'NOT GUILTY', which may or may not refer to the O. J. Simpson trial. Before the digital reissue you could only buy *Eccojams* on tape from Discogs, where albums sold for $400 each. The tracks are unnamed, discernible only as 'A1' up to 'A8', then 'B1' to 'B7'.

In YouTube comments I've seen this album described as relaxing, upbeat, addictive and, more memorably, 'like a classic rock station from hell'. To me it is far from comforting; voices are slurred to a sinister drawl, multiplying into drunken choirs which repeat a single line for the duration of every track. *Eccojams* mimics that familiar tendency, the process of half-forgetting songs. Samples come from 1970s, 1980s and 1990s hits; 'A1' features Toto's 'Africa', and 'A2' uses Fleetwood Mac's 'Only You'. Sound effects accelerate time's violence; they strip away meaning from these tracks, rendering blurred and ironic what was once sincere. 'B3' is perhaps the most poignant track of all; a voice urges us to carry on, wait another year, and eventually we'll find the happiness we seek. It mimics the sound of life's passing, chugging perniciously, meaninglessly along.

Chuck Person, creator of *Eccojams*, is in reality Daniel Lopatin, also known as Oneohtrix Point Never. He's a prolific solo artist, producer, musical collaborator and composer of film soundtracks, whose themes fall at an intersection between pop culture, philosophy and technological apocalypse. In a 2018 Reddit AMA, he said of *Eccojams* that 'it was a direct way of dealing with audio in a mutable, philosophical way that had very little to do with music and everything to do with FEELINGS'.[7]

Part of why fans relate to *Eccojams*, I suspect, is its oddly moral subtext. The protagonist of *Eccojams* is exhausted, yet constantly striving for a better life; hoping, having those hopes dashed and hoping again. The songs sampled – Toto's 'Africa', Ian Van Dahl's 'Castles in the Sky' and even Chris de Burgh's 'The Lady in Red' – inspire optimism, but then trick us and repeat, descending into automated mockery. Soon we're wading through sludge, through memories that have lost coherence and context. The voice tells us to hold on another year– but that year has long since passed, and we are as lonely and confused as ever.

Eccojams asks its listener to 'be real', but who can tell us what realness is? Each vaporwave album is a museum, and museums are rarely assembled without bias. These were artists, once, who put work and passion into their music; now their words gleam meaningless, polished by decades of replays.

Within vaporwave the frame of cultural reference is most often American, but that culture is also depicted as faded, tragic, and linked to the past. It's overlaid, too, with Japanese influences – Nintendo, *Sailor Moon* and *Neon Genesis Evangelion* are all referenced – recalling the 1990s obsession with all things 'big in Japan', most of all technology, and the slow death of American cultural dominance, which vaporwave implies is already long over. These songs are in mourning; their world is dead, because vaporwave itself helps to bury its influences.

In my lifetime, the music industry has passed through Napster, LimeWire, Kazaa, MySpace, iTunes, Spotify and YouTube, into a limbo where critics and record store staff are replaced by algorithms. Without knowing it we crossed over a threshold; today the thing that entertains us is watching us, harvesting data and predicting our tastes.

Perhaps music will never be innocent again. Perhaps music is an agent of surveillance. Vaporwave longs for a time when music was more important than technology; simultaneously, it mocks and reaches towards its raw material – songs from the past – certain that we'll never be capable of such sincerity again. There are no subcultures any more, only groups of sad people convening on the internet. Only young people, poised on the cusp of becoming old, inheriting a world that will likely be uninhabitable.

6. ░▒▓新しいデラックスライフ▓▒░:▣世界から解放され▣

This album is called vaporwave's darkest joke, almost as often as it is labelled unlistenable. The names of both artist and album are a collection of symbols and Japanese letters, meaning, respectively, 'New Deluxe Life', and *Liberated From the World*. The cover depicts the character Amuro Ray from *Mobile Suit Gundam*, a late-1970s anime about giant robots and interplanetary war. It opens with a grating, abrasive sample, a voice that sounds like it's saying 'URAP URAP URAP URAP URAP'. This is repeated, perhaps hundreds of times, like a machine-brain slowing down in the moments before its death. Time no longer exists; instead we have self-replicating sadness, sadness by design.

Every year, someone on Reddit asks if vaporwave is dead. The question has been repeated so often that it has become something of a meme. You get the sense people want it to die, if only so that it can be something we're nostalgic for.

I don't believe vaporwave is dead, yet here is its zombie. *Liberated From the World* raises questions about what vaporwave really is; does it require a melody? Is it defined by its intention, or by its sources? The pink light is gone – *all* light is gone, as though this album comes from the deep web and will give you viruses.

Its name is almost ungoogleable; *Liberated From the World* lures us beyond the filter bubble, into vaporhorror. A comment on YouTube compares it to the viral video from the horror film *Ring*.

There are human voices here, but they're faded like ghosts. This is the sound of machines inheriting the earth, a sonic dramatisation of living with – and slowing being killed by – technology. This is the sound of synthetic rumination, samples of old game shows, and a haunted ballroom filled with broken automata.

The world vaporwave speaks to is that of the cyborg; a hive mind bombarded with commercial thoughts, and a body isolated and dulled, slouched in a chair or lurching between cash registers. Nostalgia is a trap, almost as much of a trap as technology. What was I sleepwalking towards? Then it loops back – 'URAP URAP URAP' – and repeats, as though we're trapped in hell.

PART 2
Battery Life

Monstrous Energy

I HAVE ON MY DESK a medically inadvisable amount of Monster Energy. The five 500ml cans in front of me contain a total of 758 mg of liquid caffeine, and 1,700 per cent of my recommended daily dose of vitamin B12. They are branded with a large, lurid 'M' that looks as though it was etched into the can by some three-clawed animal. The text on the side of each can, printed in neon green and silver grunge fonts, encourages me to 'unleash the beast'.

Reader, even for you I am not going to subject my internal organs to all 2.5 litres. But I will sample each flavour – Original, Ultra, 'The Doctor', Rehab and Absolutely Zero – moving them contemplatively around my mouth, the way I imagine wine critics do.

I begin with Monster Original. It opens with a kick of carbonation, then gives way to the playground flavour of bubblegum cut with canned cherries and an afterthought of Lemon Pledge. Ultra is cooler – glacial, almost – and conjures thoughts of a more medicinal 7up, or perhaps Prosecco. I wonder briefly if I can discern a murmur of Panax ginseng root, or guarana extract – both of which appear in the list of ingredients – but I have tried enough herbal sleeping aids to know that this taste is not of the natural world. I move on to 'The Doctor', named in reference to the nickname of Italian motorcycle racer Valentino Rossi. It's the most explicitly citrus-flavoured of the lot, and tastes the most like a traditional soft drink: like lemon drops, with a note of creamy orange.

Monster Rehab, the lemonade variety, was mysteriously retired from the Irish market in 2017. I know this because for years it was my favourite Monster, one I would buy on nights when I was working to a deadline. I liked it because it was smoother and easier to drink than the others – it was non-carbonated, mild and semi-sweet, a good gateway variety for novices. The 'Rehab' name hinted at benevolence, as did the inclusion of niacin, riboflavin and green tea among its ingredients. I enjoyed the tension between its curative aspirations and Monster's 'extreme' branding; apparently at once stimulating and soothing, 'Rehab' felt like an energy drink masquerading as a health food, a cure for the hangover that is life.

At some point, however, my favourite Monster quietly disappeared from supermarket shelves. The can currently on my desk is an American import, 'Monster Rehab Raspberry Tea', which I found in a fridge at the back of a shop selling supplements to bodybuilders. Its taste is disappointing: a homeopathic shadow of raspberry, tannic undertones and the blandness of apple juice. I cannot detect coconut water or goji berry, though they're listed on the side of the can. I wonder if I have idealised Monster Rehab in memory: perhaps it always tasted like a mixture of Buckfast and Bronchostop.

After the tasting, my teeth feel gritty. I begin to crave carrot sticks, and milk, and all that is wholesome. I think of my dentist, and am for several seconds overwhelmed with shame. My hands shake. A headache is beginning to pierce the front of my head. There are friends of mine who tell me they drink multiple Monsters before breakfast, and I have read stories about an American football coach, a burly man named Ed Orgeron, known for his Cajun-accented bellowing at the sidelines, who consumes ten cans of Monster per day.

I am no Ed Orgeron – I draw the line at two cans per day – but I share with him, and other devotees, the knowledge that the taste of Monster Energy is ultimately irrelevant. There is no audience for tasting notes, because the varieties all end up tasting the same. They taste of Monster, the way Coca-Cola can only really be said to taste of itself. They're fizzy, so fizzy they sometimes feel like they'll cut open your oesophagus on the way down. They taste like the memory of lemons, like bubblegum after it's been chewed for half an hour. The comparison fans most often make online is with Calpol, the children's painkiller, and this is accurate; Monster brings to mind sick days, and the mildly deviant feeling of being at home when you would usually be in school.

What does Monster Energy want from us? Is its name a warning? Its promise is that it will accelerate and transform you into a newly Monstrous self, a self more energetic and reckless, living a life most of us are already too old for. In this sense, it is a lot like the non-FDA-approved herbal boner pills sold at gas stations in America: it is consumed in haste, by desperate people, who don't dare to read the packaging too closely. Its appeal transcends its function as refreshment, or even as a source of sugar and caffeine and taurine and 220 mostly empty calories. Monster speaks to a world in which we require super-normal energy inputs to survive an ever-accelerating cycle of work and consumption, work and consumption, work and consumption.

Before Monster, my energy drink was coffee. When I was at university in England I was severely anorexic, and drank instant coffee to buoy myself up during the day instead of eating. Sainsbury's Rich Roast – cheap granules with a carbonised smell – cost £2 a jar. I didn't think much about this habit until one day I noticed I had ringing in my ears, a tuneless backing track to the sound of my heart aggressively thumping. I realised I was feverish, and the amount of air I was taking into my lungs suddenly seemed insufficient. As I lay down on my bed, I realised I'd drunk twelve cups of coffee that day, and had eaten only an orange.

By third year I was doing better, and eating enough to pass as normal. Energy drinks found me in time for my final exams. One afternoon in late May, a crate of cans was dropped off at the porter's lodge by a rep from Relentless, an energy drink owned by the Coca-Cola company and marketed with the slogan 'No Half Measures' – a reference to its cans being twice the size of Red Bull's. Their branding was vaguely ecclesiastical, with Blackletter typeface. Along with 'Inferno', the original, other varieties included 'Devotion', 'Immortus' and, incongruously, 'Mango Ultra'.

I went to an old-fashioned college – Magdalene, pronounced 'maudlin' – at the University of Cambridge. Magdalene was a beautiful but strange place. The last Cambridge college to admit women, in 1988 (some students wore black armbands in protest), it is home to a gargoyle that spits water into the River Cam, made in the image of an Elizabethan banker who mismanaged the college's property. There's an enduring myth that a team of rowers from St John's, the bigger, richer college up the road, once fixed a sharpened spear to their boat and murdered one of the Magdalene crew's coxes.

At Magdalene we had a twenty-four-hour library, a narrow, meandering space at the back of the college filled with ephemera including a note written to T. S. Eliot by his wife, instructing him to put the milk back in the fridge (it has the addendum 'Done', written by the man himself). This was where I settled, with my books and my laptop with its broken, hissing fan and my can of Relentless, which I now noticed had several lines of poetry written on its side:

But I have lived, and have not lived in vain;
My mind may lose its force, my blood its fire,
And my frame perish even in conquering pain;
But there is that within me which shall tire
Torture and Time, and breathe when I expire.

Above was written, in capital letters, 'ENDURE ETERNALLY'.

I would have known, had I done enough reading, that these lines were from *Childe Harold's Pilgrimage* by Byron. I would also have known, had I *really* done my reading, that 'ENDURE ETERNALLY' came from Dante, from the inscription written on the gates of Hell.

That week, and in the days leading up to the exams, I bought Relentless again and again, and returned each night to the college library. I was convinced that everyone else had an edge, somehow, whether it was bought from 'tutoring' websites or ingested in the form of caffeine pills or ADHD medication. Maybe this could be my edge.

Life at Cambridge, especially during the college terms, is known to students and teachers as 'the bubble', a collective delusion broken only when leaving the city (this threshold is understood to be represented by a lamppost on the green at Parker's Piece known as 'Reality Checkpoint'). As the exams drew closer I saw my classmates less and less, and our shared reality began to fray at the edges. One of my classmates dropped out. Rumours circulated of a mathematician at another college who locked herself in her room for three days then ran into the road and lay down in the middle of traffic. One night I found the library empty apart from a girl seated at one of the tables, and as I passed her I heard the sound of rustling. I looked down and noticed that the lower half of her body was encased in a sleeping bag. Later I heard talk that a group of students had also tried to install a kettle in the library, but that it had been confiscated by college staff. Finally on exam day, a friend turned up with her hand wrapped up in a sling; she had taken too many notes, and now her wrist was broken.

By the time the exams were over and the results displayed – in keeping with sadistic tradition – on the wall of the Senate House, I couldn't wait to leave Cambridge. What was it for? All this work, this frenzied dedication to writing from centuries before. Watching friends lose their minds (as I had lost mine – I'd come home after my first term weighing no more than six stone), it struck me as a culture of needless extremity, one that I assumed, for whatever reason, could not possibly be matched by the world outside.

At some point, Monster, whose garish cans I noticed on other people's desks in the Magdalene library, became my energy drink of choice, and I have stuck with it ever since. Even now, the purchase – often made after dark – feels vaguely illicit, like I am buying something dangerous, a product certainly not meant for girls. I have become a connoisseur of rare Monsters, and of €3.50-for-two deals at my local garage. But the frisson I get every time I buy a can of Monster can be dismissed as a triumph of marketing, because beyond their high-concept names and hysterical rhetoric, there's very little difference between the ingredients of the leading energy drinks. They almost always contain an amino acid called taurine, a number of B vitamins, a fruit-like flavouring agent and varyingly 'extreme' amounts of caffeine.

For all the dystopian novelty of their branding, energy drinks are not an entirely new idea. These products, in various guises, have been around for centuries. Their history is related to the history of shifting cultural attitudes to work, the body and its limits. Over the years, energy drinks have swung between deviance and legality, recovery aid and coping mechanism, narcotic secret and commercialised aid to working life.

The inventor of Coca-Cola was a morphine addict; it is said

that he was looking for a safer alternative to his drug of choice. He had his first commercial success with a product called French Wine Coca, a cocktail of wine, cocaine and caffeine that was conceived and marketed as a 'nerve tonic'. Coca-Cola, launched in 1886, was an alcohol-free variant, but cocaine remained in the mix for a number of years. By 1902 the company switched to using 'spent' coca leaves, and the product was fully 'decocainised' soon after. In 1911, the US government sued Coca-Cola under the Pure Food and Drug Act, arguing that the drink's caffeine content was damaging to consumers' health, and that its ability to disguise fatigue could lead to fatal levels of exhaustion. The case went all the way to the Supreme Court, which ruled in the government's favour in 1916. The cocaine was gone, and the caffeine was reduced, but Coca-Cola continued to market itself to office and factory workers as a pick-me-up.[1]

In the decades that followed, energy drinks cultivated an association with productivity, rather than fun. In the UK, Lucozade began as Glucozade, a drink launched in 1927 to 'replace lost energy' in people recovering from illnesses.[2] In Japan energy drinks became part of the post-war economic boom; Lipovitan-D, a vitamin-fortified tonic sold in tiny, medicinal brown bottles (still available today), was for decades marketed to salarymen with TV spots starring Arnold Schwarzenegger as a harried office worker. It sold 100 million bottles per year in 1965, 200 million in 1970, and 400 million per year by 1980.[3] Another product, Suntory's Regain Energy Drink, has TV ads featuring an army of men in suits scaling a skyscraper in order to get to work on time. In another of their ads, a superhero in a suit asks 'Can you fight for twenty-four hours a day?' Later, he chants, 'Businessman! Businessman! Japanese businessman!'

Energy drinks found a ready market in the military, another historically sleep-deprived group. The history of wartime stimulant use is well-documented – during the First World War, between morphine and vodka shots, soldiers dosed themselves with cocaine pills marketed under the name 'Forced March', and by the Second World War, both Axis and Allied forces preferred amphetamines; Nazi soldiers consumed an early form of crystal meth, sold as 'Pervitin', as well as speed-laced chocolates and cocaine chewing gum, while the British and American forces opted for Benzedrine tablets. Japanese soldiers, meanwhile, favoured a methamphetamine product called Philopon, its name derived from the Greek *philoponus*, meaning 'he who loves labour'.

This practice overlapped with a dependence on energy drinks; during the First World War, the US Army built coffee roasting plants in France to keep the front lines well-stocked, and bought roughly 37,000 pounds of instant coffee per day (itself a relatively new invention – sales took off after the war, when returning soldiers began to ask for it at home). More recently, the military has preferred obscure energy drinks, including 'Wild Tiger', also known as 'Adderall in a can', a favourite with American soldiers in Iraq, and 'Rip It', provided for free at military commissaries in Afghanistan.

Today the world's best-selling energy drink is a Western version of a product from Thailand; in 1984, Austrian marketer Dietrich Mateschitz brought the Thai energy drink Krating Daeng to Europe, reformulating it, and rebranding it as Red Bull. The original drink was popular with truckers and labourers, but Red Bull was marketed as edgy and fashionable. Mateschitz paid DJs and students around Europe to serve the drink at parties – a 'word of mouth' campaign, much like how Relentless targeted my college with free samples during exam season.

Today Red Bull is the global market leader, though Monster is not far behind, leading a pack of challengers in an ongoing battle to raise the bar of high-octane tastelessness and brazen trolling. (For the latter, consider the drink launched in 2016 by two Polish businessmen in Greater Manchester: Brexit.)

Over the years I have compiled a taxonomy of energy-drink brand varietals:

Sex	Pussy, Beaver Bomb, Bacchus-F, Bawls, Bang
Violence	Semtex, Bomb, Monster Assault, Monster M-80, Monster M3, M-150, Monster Punch, Monster Ripper
Animals	Red Bull, Shark Stimulation, Sting, Bulldog, BeeBad, Power Horse, Little Dragon, UberMonster
Machines	Monster Unleaded, AMP Energy, Reload, Sparks, Uptime
Morality	Burn, Hell, Inferno, Red Devil, Aspire, No Fear, Virtue, Mother
Medicine	Dr Enuf, 'The Doctor', Monster Rehab
Power	Street King, Rockstar Revolt, Brexit, Pimp Juice
Other	Cocaine, Hype Energy, Urge Intense, Monster Khaos, Relentless, Loco, Zombie

Though I have never tried it (apparently it's only available in Northeast Tennessee), Dr Enuf might be my favourite energy drink name, because it speaks directly to the question energy drinks raise: whether any amount of energy, or work, or consumption will ever be enough.

Energy drinks traffic in the language of satisfaction – what Coca-Cola's marketing once called 'the pause that refreshes' – even as they gear us up for further exhaustion ('Reload'). We consume them so that we can consume more ('Cocaine'), refuelling the body as we would a machine ('Monster Unleaded'). These products are symptomatic of a world where the language of the 'extreme' has seeped into everyday life. Where once energy drinks were geared towards recovery and wellness, now they are about performance enhancement in the capitalist hellscape.

My five-variety tasting was but a tiny sampling of the available Monster variants, which are numerous enough to populate entire supermarket fridges. Some of these can only be viewed as marketing jokes, products of a commercial culture so bloated as to revel in its own excess. Monster Tour Water, for example, is just water; it is said to have been created for roadies on the Warped rock tour after they complained that they could not drink *only* Monster Energy in the summer heat.

The online crafts marketplace Etsy is populated with homemade Monster Energy tributes. A cursory search reveals a host of products for sale, including claw-mark vinyl car decals for €4.53, and a Monster-printed scrub cap, €13.62, meant for wear by nurses and doctors. I have also found repurposed ring-pull earrings, a five-foot Monster wall decal, Christmas tree ornaments, and a Monster Jesus trucker cap, where the green 'M' has been replaced with crucifixes. These fan-crafts are naive art next to the official Monster merch, but not everyone can afford, or pull off, official Monster Team gear such as the Thor Monster Pro Circuit 2013 Phase MX Pant ($136.95, a must for professional stunt riders) and the Hoonigan Monster Black Ken Block Racing Division Team Mechanic Tank Top ($41.99).

Monster Energy's marketing copy seems to be based on a child's idea of how a drug dealer would talk ('We went down the lab and cooked up a double shot of our killer energy brew'). This crude pharmaceutical glamour speaks to the part of me that's hopeful and broken, the part of my personality that leads me to try – and subsequently abandon – a new antidepressant at a rate of roughly one a year.

In April 2018 I was sitting at home watching a film with my soon-to-be-ex-boyfriend, drinking a can of Monster Energy Zero Ultra. (It was once the case that I was too embarrassed to consume the stuff in front of boyfriends; I fixed this by only dating men who drink it too.) This particular boyfriend shared my fascination with gaudy machismo; looking back, I think our mutual Monster enthusiasm was one of the foundations of our relationship. I told him, once or twice, about my intention to one day write a Grand Unifying Theory of Monster Energy, a piece of sprawling, half-baked cultural theory aiming to make sense of the drink's place in our lives.

That night, he had come over to watch Alex Garland's *Annihilation*. In it, a team of scientists travel into a zone known as 'the Shimmer', a mutated world where the cells of different organisms commingle and evolve into something rich and strange. An alligator has shark's teeth, flowers bloom from the antlers of deer, and different varieties of plant grow from the same stem. As we watched, I noticed that my not-yet-ex was frequently checking his phone. I didn't mention it, because I myself find it hard to sit through films, sometimes, without doing the same.

The scientists themselves began to mutate. One transformed into a plant, another became part of the bear that devoured her. They'd become a new kind of monster; the horror of the film stemmed from its promise of life beyond death, of metamorphosis. My not-yet-ex-boyfriend was checking his phone again. A tweet he had posted that afternoon had gone viral: it had over 11,000 retweets. He checked again: over 30,000 'likes'.

I took out my own phone. My not-yet-ex-boyfriend had tweeted a picture of the side of a can of Monster Energy, the 'Assault' variety, which contained the following copy: 'At Monster we don't get too hung up on politics ... We put the "camo" pattern on our new Monster Assault can because we think it looks cool. Plus it helps fire us up to fight the big multinational companies who dominate the beverage business.' It signed off with an urge to 'Declare war on the ordinary!', and the slogan 'Viva La Revolution!'

My not-yet-ex tweeted, 'what stage of capitalism is this', a play on the Twitter meme where the phrase is paired with dystopian news headlines, retweets from politicians and other updates from the frontlines of grim modernity. As the night went on, the 'likes' reached over 63,000, alongside 16,000 retweets. Within hours, the website *Business Insider* wrote it up as a clickbait news story. It was an extreme response, befitting its subject. It confirmed that others saw something strange, and even foreboding, in Monster's rhetoric.

I wanted to be happy for my not-yet-ex's success. He deserved recognition. He deserved followers. But all I could feel was jealousy: he'd distilled my idea down to its most accessible form. I wondered where this sense of competition came from, this instability, and bitterness, and fear.

I tell myself that I need Monster Energy to finish a deadline. It takes away doubt. It makes reality sparkle. Starry-eyed, shaking, my mind becomes a hive of bees, my heartbeat an ill-mannered rattle. I want to write everything at once, my hands typing almost as fast as my thoughts. 'Tear into a can of the meanest energy drink on the planet'; Monster Energy puts you in a mindset where the text on the Monster can makes sense.

The more I think about, and consume, Monster Energy, the more I notice its language in other spheres of everyday life. I see the mark of its neon claws on culture, the summation of centuries of fear and machismo, and men seeking a cure for their own humanity. Monster Energy gives form to questions explored in horror and science fiction: the threat to the human body posed by urbanisation, industry, technology and capitalism. It conflates 'monstrousness' with the will to carry on.

Rumoured to have been written in a cocaine-fuelled three-day burst, Robert Louis Stevenson's *The Strange Case of Dr Jekyll and Mr Hyde* (1886) explores the effects of a serum that gives its creator a jolt of energy. The novel outlines a pattern of temptation and regret:

> I purchased at once, from a firm of wholesale chemists, a large quantity of a particular salt which I knew, from my experiments, to be the last ingredient required; and late one accursed night, I compounded the elements, watched them boil and smoke together in the glass, and when the ebullition had subsided, with a strong glow of courage, drank off the potion.

Dr Jekyll's 'transcendental medicine' causes him to transform into his alter ego, Mr Hyde, a man-beast with a skittish manner and a taste for sudden violence. Hyde is smaller and younger than Jekyll, shrunken inside his expensive clothing. He moves through the fog of Soho by night on missions of unspecified depravity. In the beginning Jekyll can switch Hyde on and off with the potion, like one of Rossum's Universal Robots, or their mythological predecessor, the golem. In this sense Hyde is the perfect worker: fast-moving and powerful, unambitious and gifted with limitless energy.

'Well, it was this way,' begins Enfield, a friend of Jekyll's, narrating the first sighting of Mr Hyde:

> I was coming home from some place at the end of the world, about three o'clock of a black winter morning, and my way lay through a part of town where there was literally nothing to be seen but lamps. Street after street, and all the folks asleep – street after street, all lighted up as if for a procession and all as empty as a church – till at last I got into that state of mind when a man listens and listens and begins to long for the sight of a policeman.

'Some place at the end of the world'; the preferred habitat of the urban monster. Moving most often by night, Hyde's 'energy' is mentioned on several occasions: we see him pacing unnervingly back and forth, gnashing his teeth. Stevenson writes that Jekyll 'thought of Hyde, for all his energy of life, as of something not only hellish but inorganic'.

The Strange Case of Dr Jekyll and Mr Hyde is set in a London that was on the brink of electrification. In the same year it was published, Karl Benz patented the world's first automobile, Jacob's Pharmacy in Atlanta sold its first bottle of cocaine-laced Coca-Cola, and Sigmund Freud opened his practice in Vienna. The following year, the word 'calorie' appeared in a series of articles in the magazine *Century*, written by Wilbur Olin Atwater, an American chemist known for his work in the field of nutrition. Decades earlier, the term had been used in lectures on heat engines by French physicist Nicolas Clément; now, as the word 'calorie' entered popular usage, people talked about the body the way they had once talked about machines.

The unsleeping, amoral Hyde is in fact more useful to the new London – then emerging – than his progenitor, the mild-mannered, patrician Jekyll. Their struggle, then, concerns the anxieties of an age in which the human body competes with machines. It's also worth noting that when Hyde sleeps he ceases to exist, and transforms back into Jekyll. For Hyde, then, it's better not to sleep at all.

Sometimes I still place hope in Monster Energy. In certain moods, 'unleashing the beast' can feel like a form of self-improvement.

'Monster is a lifestyle in a can!' But Monster is also the route to anxiety, to unfocused thoughts and unfinished sentences. It makes the screen too bright. It gives me a headache. Then the feeling sets in; nausea, like a claw reaching down my throat, into hollowness.

I think of Monster, at these times, as the descendant of another fictional elixir, one charged with hope and hyperbole and fear: the eponymous 'Ubik' of Philip K. Dick's 1969 novel. In *Ubik*, immortality is a commercial product. People survive death by entering 'half-life', their brains preserved while their bodies are frozen. Ubik is used as a spray, a hair product, a balm, an elixir or a powder. It staves off decay, reverses entropy, and preserves the present long enough to enjoy it, although it cannot do so for ever.

Ubik's hero, Joe Chip, is forever on the verge of being overwhelmed by debt, by paranoia and by the appliances in his kitchen, which demand payment in order to work. (In one well-known scene, a tiny robotic voice inside his front door threatens to sue Chip as he jams it open.) Machines compete with an increasingly helpless human race whose use of Ubik locks them into a cycle of dependency, swinging between decay and ghoulish temporary vigour. 'A dismal alchemy controlled him,' Dick writes of Joe Chip, 'culminating in the grave.'

Like the copy written on cans of Monster Energy, Ubik's messaging betrays a number of human anxieties, even as it heralds their cure. My favourite Ubik ad is a TV spot, perhaps the novel's most lyrical and most transparently deranged:

> One invisible puff-puff whisk of economically priced Ubik banishes compulsive obsessive fears that the entire world is turning into clotted milk, worn-out tape recorders and obsolete iron-cage elevators, plus other, further, as-yet-unglimpsed manifestations of decay ...[4]

Likewise with Monster Energy; the more we consume, the more we need it, the more it begins to restructure reality on its own hyperbolic terms.

Of course, I crash after drinking Monster Energy. I give up on my deadlines and set an early alarm. I brush my teeth and stare regretfully into the mirror, examining my jawline for Monster-induced acne. Sometimes, I tell myself I'll never drink Monster again.

There's something grim about the empty Monster Energy cans in the bin beside my desk. They speak of a dependence, not on some glamorous illicit substance, but on the rough product of centuries of fragile masculinity left unchecked, fermenting in a hyperbolic brew.

I consider the future of energy drinks, which will involve dependence on a single drink not only for energy, but for all of the body's sustenance. This theory is possibly the result of too much Monster, typed out by shaky, caffeinated hands. But the makings of this dystopian scenario have already been brought to life.

In 2013, a 23-year-old software engineer called Rob Rhinehart posted a blog titled 'How I Stopped Eating Food'.

> In my own life I resented the time, money, and effort the purchase, preparation, consumption, and clean-up of food was consuming. I am pretty young, generally in good health, and remain physically and mentally active. I don't want to lose weight. I want to maintain it and spend less energy getting energy.[5]

The post goes on to describe an experiment lasting for thirty days; Rhinehart creates a nutritionally 'complete' powdered food which can be prepared with water, and stops eating conventional, solid food. He becomes more efficient, saving both time and money. His invention, Soylent, takes its flamboyantly dystopian name from Harry Harrison's 1966 novel *Make Room! Make Room!*, in which citizens of an overcrowded near-future New York are fed on rations of Soylent, a mixture of soy and lentils. In Richard Fleischer's 1973 film, *Soylent Green*, Soylent famously turns out to be made from people.

Over one month consuming Soylent (the non-cannibalistic kind), and nothing else, Rhinehart spends a total of $50 on sustenance. His skin clears up, he sleeps better and he stops defecating, as he is consuming only a few grams of fibre per week.

He writes, 'I haven't eaten a bite of food in thirty days, and it's changed my life.'

As the post continues, Rhinehart evolves a kind of anti-culinary ideology. Using his own results as proof, he argues that the world would function more efficiently, on a macro and micro level, if more people consumed Soylent. He writes with a glassy-eyed fervour: 'After a week advertisements for fast food looked repulsive. All I crave is Soylent.' He says that his focus and his reflexes have improved, and describes how Soylent has gifted him a heightened perception, a sensory vividness which reads like a cross between an Erowid trip report and a sales pitch for nootropic stacks ('I find music more enjoyable. I notice beauty and art around me that I never did before. The people around me seem sluggish.').

Rhinehart's idea – and the names of Soylent and its coffee-infused version, Coffiest – comes directly from sci-fi novels of the 1950s and 1960s that posited worlds in which food was viewed solely in terms of caloric potential, and hence the ability to fuel work. While initially ridiculed, Soylent remains on the market with a cult following branching beyond Silicon Valley. There are now multiple flavours of Soylent, as well as rival products with names like 'Huel', 'Powder Matter', 'Queal', 'Mana', 'Bertrand' and 'Joylent'. On a recent trip to London I noticed Soylent stocked in fridges at newsagents, and advertised on billboards at train stations; the brand had signed a distribution deal with WHSmith.

Rhinehart comes from a tech background, and Soylent's early backers included Google Ventures, Andreessen Horowitz and Elon Musk. To me it seems obvious that Soylent was created by and for the tech industry. It speaks to a customer willing to sacrifice pleasure for convenience, variety for routine, sociability for solitude; the kind of customer who is chained to their laptop, and accustomed to working in sprints.

Soylent is the inheritor of energy drinks and narcotic elixirs.

It articulates a view of the body as a testing ground, for products that can help it compete with machines. Food becomes fuel, nothing more, for the body's battery; Soylent's marketing carries an economic subtext, a portrait of a worker who is starved for time, obsessively focused, and prone to periods of dangerous self-deprivation. Work, to the Soylent consumer, is as potent, as hypnotic and intoxicating as recreational drugs are to other young people.

A study conducted by the Mayo Clinic in 2015 found that energy drinks raise levels of the stress hormone norepinephrine by just under 74 per cent.[6] Every year, news stories appear about health risks, and occasionally deaths, linked with their consumption. They make for freakishly compelling headlines ('Man turns yellow and develops severe hepatitis after drinking five cans of energy drink a day', 'World's fattest man dead: Andres Moreno dies "after energy drink binge" aged thirty-eight'), but these stories appear with such regularity that they have become background noise, perhaps even contributing to the products' appeal.

A study published in 2019 in the *Journal of the American Heart Association* indicated that energy drinks cause more harmful changes to blood pressure and heart function than caffeine alone.[7] Apparently they make the blood thicker, and can send people into cardiac arrest, and are linked with muscle twitching, restlessness and bouts of anxiety. According to Jamie Oliver, they are 'turning our kids into addicts'.[8]

What am I drinking Monster Energy for? I'm looking for refreshment, and clarity of mind, yet when I think about it, Monster Energy very rarely delivers either. It does give me confidence – or perhaps only energy, masquerading as confidence – enough to make my fingers dance over the laptop keys. I feel as though I don't really appreciate the recreational possibilities of substances like speed and cocaine. The latter especially, in the past, seemed to barely affect me; it only made me more decisive, and calm. On speed I enjoyed reading a book. On Ritalin I fell asleep. It seems, by my own unscientific methods at least, that Monster has inured me to the effects of other stimulants.

Before writing, I start with a deadline and a word count. I have raw material: recorded interviews, notes from articles and books, and the dazzling, apparently limitless expanse of the internet. Every time, almost without variation, I experience the same sequence of thoughts. I try to type, but tell myself I'm tired, not physically, but mentally and emotionally. I believe I'm not capable of carrying out in writing the ideas I myself have conceived of. I am not good enough to do my own work. I remedy this with Monster Energy, and occasionally Valium.

Increasingly I have these same thoughts when I think about the future. I wonder if I will be replaced by a machine in my lifetime, and whether it will be before or after I pay off my student loan. I think about how I spend my time and energy, about how I can live meaningfully. It feels in these moments like the answers to my questions are circling around my head, just out of reach.

Ultimately, Monster Energy only makes these feelings worse. It hurls its consumer into the future, automating the body while the mind falls into a stupor. It is the descendant of the stimulants fed to young soldiers during the world wars, the quack cure for modernity, the simulated enthusiasm-on-tap for jobs that pretend to be meaningful. It is fuel for accelerationism, the philosophy outlined by Nick Srnicek and Alex Williams, in which, in the face of apocalypse, we embrace rather than resist radical change.[9] It is a taste of the mooted Chthulucene epoch,[10] in which our bodies will cease to function without commercial stimulation. It is our mortal attempt to add hours to the day, to bolster the body against its own will to give up. Against the city, against machines, and against nature, which feels more alien to us every day.

All Watched Over 1: Always On

IN DECEMBER 2018, I went to London for a month to remember how to write. Instead I forgot how to sleep.

How to sleep? The internet is full of advice from strangers. Lie with one leg hanging off the side of the bed. Drink herbal tea. Most importantly, switch off your computer. Insomnia is where our latent technophobia finds expression, coupled with our blatant technological dependence. Headlines from recent years state that blue light from LED screens is disrupting circadian rhythms, and that teenagers wake to check their social media feeds in the night. Social media use before bedtime is said to decrease your chances of a good night's sleep, and heavy users are twice as likely to suffer from sleep disorders. Separate studies correlate depression with tweeting between 9 p.m. and 6 a.m. – 'healthy' Twitter users are more likely to post during the day. And all of these issues, of course, are aggravated by modernity, because silence is impossible, rest is impossible, in a world that's always on.

We will sleep no more, not properly. These problems are discussed online, where communities of believers, sceptics and insomniacs congregate to feed an ongoing 'sleep crisis'. Even in sleep we are dependent on the thing that disrupts us, because the cure for insomnia is out there, somewhere on the internet.

Perhaps technology has triggered my insomnia. Since becoming a freelancer, the boundaries between work and leisure in my life have collapsed, as have the borders between waking and sleeping. Sleep has long served as an indicator of the state of my mental health; under stress I stop sleeping, or experience strange dreams or sleep paralysis. Back in 2016, as depression took hold, I started going to the Night Gym. Now it's the darkest part of winter, and once again, I no longer observe the social contract with daylight. If there's a deadline, I stay up all night. If I'm anxious, which I am a lot of the time, I stay up all night anyway.

Most days I work and sleep in the same location: sitting on my bed, with my laptop in front of me, very often while wearing pyjamas. The internet has picked up on it; in sidebars and sponsored social media posts, I get ads for luxurious bed linen, silk pyjamas sets that I cannot afford, and occasionally weighted blankets, a therapeutic tool used to treat anxiety, which I take as online marketing's expression of concern that I will die alone.

If I'm feeling energetic then I sit cross-legged on top of the covers, stooped over the screen. Nearing nightfall I lie back and type lazily, the machine across my ribcage, the LCD screen bearing down on me. As I drift into sleep I cannot let go; I open Netflix in a tab and watch *It's Always Sunny in Philadelphia*. I cannot go to sleep without background noise. I don't want to be alone; I'm afraid of the sounds I'll find if I grant my mind silence.

I have dwelled for years on the borders of insomnia, but throughout December it creeps up on me in a more decisive form. Slowly I withdraw from sleep as though detoxing from a drug. During the first week I fall asleep at 1 a.m. Later it crawls forward to 2 a.m., 3 a.m., 4 a.m. Soon I am sleeping for four hours per night, then three, then two. For three days I don't really sleep at all. Instead I cultivate stasis; I lie in bed, the laptop propped across my solar plexus. I used to envy people who lived like this; I thought they had cheated daylight and gained double the time for productivity. But now I'm apathetic, blunt, incapable of anything but watching the same episodes of the same show that I've fallen asleep to hundreds of times before.

London is an unsleeping city, but London is not the reason I've stopped sleeping. Something has shifted; I'm overwhelmed by grandiose concepts and tiny details I can't stop from reverberating around my mind. Night brings a certain kind of anxiety, one that sleep might cure, if only it were possible to achieve it. Night is when my memory turns against me; the past flashes on my mind, and shame keeps me awake – shame, and the suspicion that everyone hates me.

There's a term that comes to mind here, one that feels oddly appropriate to my circumstances: 'woke'. 'Woke' has a complicated history. It originally meant staying aware of social problems and injustices, and never letting one's guard down. But more recently it has come to mean something more like complacency, a virtue which is self-declared rather than earned. The phrase 'stay woke' appeared in a 1938 song by Lead Belly, and resurfaced in a 2008 song by Erykah Badu called 'Master Teacher'. More recently, it was popularised as a phrase by Black Twitter. But wokeness became the stuff of parody the minute it fell into the hands of white people, culminating in the appearance of Jack Dorsey, the CEO of Twitter – a company with, it was noted at the time, a 3 per cent black and Latino workforce – at the 2016 Code Conference, wearing a grey T-shirt emblazoned with the Twitter logo and the phrase '#StayWoke'. This should have stopped 'woke', as a trend, in its tracks, but somehow it didn't, and the term was adopted into the lexicon of online life.

I don't particularly aspire to wokeness; it's a losing game, one where someone will always find something about you to object to if they want. But that doesn't stop me from self-interrogating, nor does it help me to sleep at night.

Now I start checking up on people: rivals, romantic prospects, exes, disembodied Twitter avatars I long to impress. The internet is more limitless than ever, and I am more distractible. I start to think of myself as a terrible person. I nurture the idea that certain people despise me, are out to get me, and that I deserve it.

What torments me most is a sense of the internet as an archive, one that records all my past mistakes. I download all my past tweets and go back through them, month by month, year by year, deleting anything which might read as offensive. I contemplate deleting my account entirely, but find that I can't bring myself to do it. Then I revisit other old accounts; Reddit upvotes, unflattering Instagram pictures, things I might once have Liked but which I don't like any more. Each night brings new self-edits, unfollows and purges. The internet is unending, and so, apparently, is the trail I've left behind.

At times I believe that I am evolving against my will, attaining the state the internet has long prodded us towards: eternal vigilance, instability, consumption. Like a battery animal, glutted on information, I am static and manic at once, seeking the scrolling feed's impossible endpoint. I get absorbed in the screen: sometimes I don't notice my heart thumping, and at others it feels like it's bursting from my chest, like I'm dying, but I can't look away.

I cannot stand to be alone, and yet I seek a cure in loneliness. For anyone with melancholy or obsessive tendencies, darkness – literal darkness, the stilling of the world – becomes a constant temptation, a drug that's available for free. You can beat the clock, and beat the limits of your own body. You can get back at your mother for what she told you growing up – to go to bed and be productive the next morning. Sleep is in my nature, but nature is optional; I'll not be bound by what nature tells me I can or cannot do.

The longer I go without sleep, the more waking life becomes like a dream, haphazard, bound by a flimsy logic where I seem to manifest in one location, then another, rather than following a linear journey between places. Voices seem louder, moving objects appear blurred. My every thought is paranoid. I don't eat much any more, although I do drink a lot of coffee. I speak and I walk as though permanently drunk and hungover at the same time.

I have come to London to write in my aunt and uncle's empty flat, but the longer I go without sleep, the more I want only to read about sleeplessness. I spend my zombie hours looking for answers online, on health websites and forums for insomniacs. I swing impulsively between craving sleep, apologising to it, begging it to take me back, and believing that I am above sleep entirely. Online I read about those who have beaten sleep, limited it, built home sleep labs for self-monitoring, supplemented sleep, or who have broken it into polyphasic pockets spread throughout the day. On my laptop each night, I journey through a conflicted cultural narrative; the history of sleeplessness, and the battle to 'fix' sleep even as we try to cure ourselves of it entirely.

Perhaps I'm trapped inside this contradiction, becoming further enmeshed with every night spent in front of a screen. I fall into a world of self-experimentation, communities and ideas aiming to control a thing scientists still struggle to understand.

The history of technology is a history of insomnia. Thomas Edison's phonograph, the first device to allow its user to easily record and play sounds on metal foil, was developed in 1877. Two years later, in 1879, Edison patented the first modern incandescent light bulb, created with the help of a team he called the 'Insomnia Squad'. Edison was famously sleep-averse; he worked late into the night with occasional power naps. He aligned his mission to popularise the light bulb with the cause of sleeping less and working more, writing in his diary:

> For myself I never found need of more than four or five hours' sleep in the twenty-four. I never dream. It's real sleep. When by chance I have taken more I wake dull and indolent. We are always hearing people talk about 'loss of sleep' as a calamity. They better call it loss of time, vitality and opportunities.[1]

Edison offered himself as proof of an overlap between technology, capitalism and the human body, a nascent biopolitics in which the former two could overcome the failings of the latter. Technology was an aid to the imagination; it brought to life the possibilities of night, unfettering the mind from dreams so that it might be turned loose on invention. To welcome technology into the home also meant accepting it into our consciousness, while awake and while sleeping.

If technology could not eradicate sleep entirely, then it could at least find ways to make those lost hours productive. Fifty years after Edison's phonograph debuted, its wayward child, the Psycho-phone, was marketed to households across America. A device for pseudoscientific 'sleep learning', the idea appeared in science fiction before and after it was marketed in real life. The Psycho-phone was sold as an 'automatic suggestion machine' intended for use while its owner slept; it was a phonograph equipped with a timer and several wax cylinders, which played subliminal learning tracks with names like 'Life Extension', 'Normal Weight' and 'Mating'.[2] An article in the *Washington Post* quotes one of its scripts: 'I desire a mate. I radiate love ... My conversation is interesting. My company is delightful. I have a strong sex appeal.'[3]

Edison invented the phonograph in 1877, and the global market for recorded music developed in the late 1910s. Over the decades these technologies were mass-marketed, and popularised, and eventually repackaged as products meant for use during sleep. Sleep was not sacred; instead it was viewed as a canvas for consumerist self-improvement. Drifting into sleep, the listener was willing to share in a commercialised dreamscape, and to trust a device like the Psycho-phone with their fragile mind (fragile, because they had turned to the Psycho-phone in the first place).

Print ads for the Psycho-phone, dating from the late 1920s, call it 'The Great Psychological Miracle'. They include testimonials from customers who credit it with improved mental health and improved finances. One person, identified only as 'H.E.S.', writes 'I now feel myself torn from the deep waters of mediocrity ... I consider it the best investment I have ever made.'

I picture the Psycho-phone's users setting the wax cylinder turning, hearing a distant voice emerge from the horn, and then drifting into sleep. What did they wish for? An ideal self, inside and out? A standardised self, vetted and perfected by quack technologists and self-help scriptwriters? Did they, perhaps, hope to be not special, but all exactly the same? Or did they hope to achieve a nocturnal community, a shared, elevated subconscious that would lead them into identical, technologically enhanced dreams? It's impossible to know, although clearly the Psycho-phone appealed to its customers' insecurities, their fear that they weren't good enough, even in sleep.

Over the decades, technology's adversarial relationship with sleep developed as computers became more accessible. During the 1990s, computer programmers would work by night because the servers used to host their product weren't powerful enough to support everyone in the office using them at the same time. Night was better, when the servers weren't under stress and could run faster. The same applied to dial-up internet in people's homes in the early 2000s; by day, the lines were busy, and the connection was slow. By waiting till night, when the average user was more likely to be alone with the computer, they could accomplish more. From its origins in the home, it seems the internet *wanted* us to be isolated, sleep-deprived and addicted to scrolling, right from the start.

I've read accounts calling coding a 'flow state'; not unlike writing, it needs to be uninterrupted. Sometimes it needs to stretch out for long periods, more than eight hours at a time. Today programmers still work late at night, or early in the morning, in order to find the chronological isolation necessary to make sense of never-ending lines of code. Night is when the notifications stop. No texts, no emails, no Slack, no Trello. You are in league with the night people, and the people they Skype with in different time zones (there is something a little self-aggrandising about this tendency, I've noticed, among the start-up people I'm friends with – sometimes I get messages from them in the early hours, where they pretend to be up late for work reasons. They'd like you to think they have someone in Palo Alto they're communicating with, when really they're just antisocial, or insomniacs, like me).

Today ordinary, non-coding people spend as much time with their computers as the people employed to build the internet. Today we sleep and wake and walk around in front of screens, not in a 'flow state', exactly, but in a submissive trance. In the 1990s and early 2000s, scientific studies revealed that the game Tetris could serve as a kind of hypnagogic imagery; spend long enough playing it, and the falling shapes begin to appear behind your eyes as you drift towards sleep. The phenomenon was named the 'Tetris effect'.

Some believe the Tetris effect corresponds to a distinct form of memory; in tests, patients suffering from anterograde amnesia, which wipes your short-term memory, still reported dreams of falling Tetris shapes after playing the game during the day, despite not being able to remember playing it.[4] Considering today's UX practices – the 'user experience', designed to make technology as seamless and addictive as possible – it seems inevitable that the websites and apps we use every day have also infiltrated our dreams.

For now, I know that my sleep is bookended by interactions with technology; the last check on emails and social accounts, before hiding under the covers, before the other side of sleep, and reaching to turn off the alarm, then checking alerts with eyes barely open, waiting for the screen to tell me what to do with my day. Technology follows me everywhere, in the daylight, in the dark, into my dreams. Like the blooding of vampires, it cannot be undone.

I remember a biohacking meet-up I attended in Dublin not long ago, out of morbid curiosity. It was a little disappointing; I watched as a series of men of varying ages delivered PowerPoint presentations and talked about ways to prolong their lives and stay thin and have full heads of hair for ever. It was only at the break that I met the kind of true eccentric I'd hoped to encounter, an older man with shaggy hair, a windbreaker jacket and indoor sunglasses with tinted lenses. His glasses flashed in the dark – the group had convened in a dimly lit underground bar in the Silicon Docklands – and while I could not see them, I was certain his eyes were full of nervous energy. With little introduction, the man with the glasses told me that blue light from laptops and other tech devices is the *real* problem of our times, a health crisis waiting to happen, and that the only way to protect oneself was to limit exposure to it by wearing special tinted glasses like his own. He described a lack of natural light in the average person's day. 'In the morning, get up and stand facing the sun for at least fifteen minutes,' he advised, before disappearing into the crowd.

He was right about sunlight; we need it for health and sanity. I, a vampire, still suffer for lack of that. The blue light debate, however, has yet to be settled; researchers have explored links between blue light exposure – especially at night, from computer screens – and an increased risk of cancer, but so far they've not found proof of any direct effect on health.[5]

This hasn't stopped opticians, health gurus and even manufacturers of 'blue light-blocking' skin creams from trying to profit from blue light panic. In response, tech companies have also moved to embrace 'dark mode'; Windows, Samsung, Apple and Android all feature an option for a darker on-screen colour scheme, apparently designed to make their products easier on the user's eyes. The other name for 'dark mode' is 'night mode'; it's been shown to save battery power (a white screen uses approximately six times as much as one with the brightness turned down all the way), but its other obvious purpose is to allow you to look at your screen longer, especially late at night or in a darkened room where the glare of a white screen would be too much.

Coders traditionally work with programmes with darker backgrounds, a tendency rooted in the past when colour monitors were rare. Today a darker colour scheme helps them to focus on the screen for longer, especially at night. Correspondingly, a page titled 'How to use Dark Mode on your Mac', published to Apple's website, hints at the real reason for dark mode's popularity; 'Dark mode makes it easier to say focused on your work, because your content stands out while darkened controls and windows recede into the background.' If you're working late into the night, your sore, exhausted eyes will try to tell you to stop working and go to sleep. Dark mode is meant for night workers and internet addicts – for people whose use of technology defies the limits of the body, and the limits of nature.

I find it funny, but not surprising, that tech companies would respond to fears about the effect of their product on their user's eyes with an option that allows them to use their devices for longer periods of time. Dark mode, to me, is a summation of how we relate to technology, swinging between suspicion and dependence. Though I have yet to try it, I am the ideal customer for dark mode: someone who uses their laptop until it hurts.

For people in America suffering from sleep disturbances, diagnosis and treatment are far from accessible. A sleep study, where doctors monitor the patient overnight for sleep disorders, costs on average $1,000 to $2,000 per night. In countries where studies are covered by a national health service, the waiting lists are long, and the results are not always conclusive. Mirroring a broader trend for self-diagnosis, online symptom checkers and dubious home remedies, the internet has evolved a sprawling community around sleep health and DIY sleep science. This uneven sleep culture has produced its heroes and villains, often self-appointed, because who can be truly certain what someone's sleep habits are? Who can know the contents of their dreams?

Arianna Huffington is the founder of the Huffington Post, once named the fifty-second most-powerful woman in the world according to *Forbes* magazine, a former conservative and a current liberal, and a campaigner for reform of popular attitudes to sleep. Huffington's book, *The Sleep Revolution*, relies heavily on a 'sleep crisis' narrative rooted in personal experience – in April 2007, Huffington fell to the floor of her $6.5 million New York apartment and her sister found her with a broken cheekbone, passed out in a pool of her own blood. Doctors tested for various causes, but eventually settled on exhaustion, and this gave Huffington a new mission: to prevent others from succumbing to the same fatigue that had briefly detained her.

The Sleep Revolution makes a case for sleep in terms of its counterpart, daylight productivity. Huffington herself is prolific, and successful, and it is assumed that her reader is – or will be – too. Chapters discuss 'mastering sleep' in terms of jet lag, 'performance enhancement' (inspired by the sleeping habits of athletes) and 'sleep in the workplace'. In interviews, the latter is especially stressed; Huffington quotes the statistic that lack of sleep costs the US economy over $63 billion per year in lost productivity (it is also mentioned that the Huffington Post's offices have nap rooms). In an address at the SAP SuccessConnect event, the name of which is enough to make me feel dizzy, she said:

> 'I'm passionate about never using the word "well-being" without using the word "productive" next to it ... We need to take well-being and wellness out of the fuzzy area of a soft benefit from the likes of benevolent HR professionals. This is hardcore. This is about the bottom line.'

Such was the success of Huffington's campaign against burnout that it inspired her to found an organisation called Thrive Global. Examining its first two or three pages of search results, it's difficult to know exactly what Thrive Global does, but it seems to involve corporate wellness programmes and a 'media platform' dedicated to the cause of promoting sleep and discouraging burnout. In a *Medium* post announcing the organisation's launch, Huffington lists Accenture, JPMorgan Chase, Uber and Airbnb among its corporate partners.

Huffington presents herself as a role model, a public commentator and wellness guru, but I have yet to meet anyone who especially wishes to emulate her. I do, however, hear people speak with a mixture of disbelief, unease and tentative admiration about her antithesis, Elon Musk. The entrepreneur and Tesla CEO has said in interviews that he slept in the office when he started his first company, a city guide software start-up called Zip2, working seven days a week. This habit has continued over the years; in June 2018, in an interview with CBS's *This Morning*, he said that he often sleeps on a conference room sofa (in response, a campaign led by YouTuber Ben Sullins raised over $18,000 to 'buy

Musk a new couch', which ended up being donated to charity). In a status update just over a week later, Musk announced that he'd be at the Tesla factory 'almost 24/7' in order to produce 5,000 Model 3s by the end of the quarter. Finally in August 2018, after several turbulent weeks including a new relationship with singer-songwriter Grimes, a feud with rapper and brujeria practitioner Azealia Banks, and a tweet about taking Tesla private with funding 'secured' from Saudi Arabia (this would later lead to Musk being sued for securities fraud and fined $20 million), Musk gave a widely reported interview to *The New York Times* in which he discussed 'the most difficult and painful year' of his career so far. Alternately laughing and breaking into tears, he described working 120-hour weeks and remaining in the factory up to four days at a time. He also mentioned using Ambien to sleep; the piece follows this with comments from unnamed Tesla board members, implying that Musk also tweets while under the drug's influence.[6]

The day after Musk's interview with *The New York Times*, Huffington published an open letter to him on the 'Stories' section of Thrive Global's website. It opens with compliments, before revealing itself as a critique:

> Working 120-hour weeks doesn't leverage your unique qualities, it wastes them. You can't simply power through – that's just not how our bodies and our brains work. Nobody knows better than you that we can't get to Mars by ignoring the laws of physics.[7]

Musk responded with a tweet, sent at 2 a.m.: 'Ford & Tesla are the only 2 American car companies to avoid bankruptcy. I just got home from the factory. You think this is an option. It is not.'

In 2016, Musk spoke openly of transforming the Tesla factory into an automated 'alien dreadnought', a move he later admitted cast the Model 3 into production hell. Admitting in an interview with Bloomberg that the drive to automate caused problems, he went on to explain:

> The reason I sleep on the floor was not because I couldn't go across the road and be at the hotel. It was because I wanted my circumstance to be worse than anyone else at the company on purpose. Like whatever pain they felt, I wanted mine to be worse.[8]

The above lines articulate Musk's philosophy on work in general. He wants to be seen to work hard. He doesn't see this as shallow or vain, because he actually will be working hard, too. But it also feels like Musk puts in these hours because he's locked in a competitive downward spiral with those he employs, and with the rest of Silicon Valley, performed with grim resolve and repetition in front of the world. In recent years, stories of overwork, injury and exhaustion among Tesla employees have repeatedly appeared in the press, along with stories about Musk himself. To Musk, work is pain. To Huffington, sleep is the salve for pain, but only in the name of more work. It's difficult to take either of them seriously, because each chooses to make their sleeping habits public for use as a marketing tool.

Each one attempts to be all things to all people; Musk seems to want to be a man of the people, but also a hero and a maverick, and a machine-like, unsleeping genius. Meanwhile Huffington's power as an advocate for sleep only comes on the back of her story about working herself to exhaustion (it would be interesting, come to think of it, to see any modern-day CEO ever admit to being lazy). They, and the world they are part of, have complicated sleep by drafting it into a broader conversation about exhaustion, business success and their status as public figures. It's hard to know if either actually sleeps well at night; they are as disingenuous as each other.

*

What a time to be awake. There is a part of me that prizes the hours I have stolen from oblivion. The need for sleep hasn't caught up with me; it has been replaced with something like a drug, a lightness of thought and a dizziness of the body. I visit my brother in South London, my first human interaction in days. In conversation I know that I'm rambling, but I laugh more, I smile, I speak loudly in declarative statements. On the train home I slouch against the wall of the carriage, suddenly dazed, and a man in a neon jacket approaches and asks me if I'm OK. I know what he's thinking. I tell him I'm tired, but not actually drunk. He nods and says, 'Oh, I thought you'd had a few.'

Another day, I take the Tube to South Kensington and visit the Natural History Museum, where there happens to be an exhibition about nocturnal animals called 'Life in the Dark'. Sleep-deprived and a little dizzy, I laugh at the synchronicity. The show includes simulations and detailed visual displays; I walk through a colony of bats conjured with sound and light effects, silhouettes flashing above me in a dim corridor. Another room mimics a cloud of bioluminescent plankton; a dark, mirrored oval room is filled with tiny colour-changing lights, which disappear then light up again in waves.

What dazzles me most is nature's fantastical symmetry; creatures have evolved unusual features to get by in darkness. I see the saucer-eyed tarsier, with the largest eyes of any mammal – bigger than its brain – and all the more capable of seeing in the dark; the flying fox, a variety of megabat found in tropical regions, which uses smell instead of echolocation to get around. Best of all are the jewelled squid and the vampire squid, crammed into tanks and lit up like stained-glass windows, glowing and translucent. They, too, have gigantic eyes, and photophores – organs that produce light, switching on and off in elaborate patterns. They are the natural template for a computer's LCD display.

In the next room, a video is projected on the wall. It's shot from the perspective of a team of scientists inside a submarine, encountering rare sea creatures for the first time. There's something about this combination – the awe and disbelief in their voices, and the slow, deliberate grace of the luminous fish – which moves me to tears. Scientists believe that cephalopods achieve REM sleep, like humans, during which their bodies twitch and change colour. Perhaps they're dreaming, but their dreams look nothing like ours.

I take comfort in the knowledge that these creatures don't care about human life; they serve as a reminder of all the things we do not know. Perhaps there are parts of human life, too, that we'll never understand completely: our consciousness, our dreams, our unreachable sleep.

Leaving the museum, I can't stop thinking about these creatures, how they have evolved in the dark while we as humans are pledged to daylight. The gloomy octopus, found in reefs around Australia, can build cities from rocks and discarded shells. But cephalopods don't have capitalism – at least, not that we know of, yet.

I still feel oddly energetic, and detached from reality. Late at night I write a vaguely unhinged email to my ex-boyfriend, enthusing about squids. I wonder if the time at which I send it will seem strange. Why do I worry? Technology never sleeps; the internet operates 24/7, and encourages you to follow its example.

Perhaps what I've appropriated, accidentally, is a routine of polyphasic sleep. The practice, popularised in online communities, involves breaking one's sleep schedule into pieces – either one 'core' sleep and several naps, or only naps, spread throughout the day – in the name of getting more time out of the standard twenty-four hours. Various polymaths are always named as polyphasic sleepers, including Leonardo da Vinci and Nikola Tesla, although nobody can find proof. The architect and inventor Buckminster Fuller followed a schedule of thirty-minute naps every six hours, for a period of two years. However he abandoned the habit when no one he worked with was willing to follow his example; they complained that he was disrupting their work day.[9]

If online communities dedicated to it are any indication, polyphasic sleeping is always meant as a productivity hack. A Kickstarter page for a 'brainwave-monitoring sleep mask' geared at polyphasic sleepers, called the NeuroOn, reads 'Enjoy longer, more efficient days. Sleep a total of just 2 hours a day and let your productivity soar.' There's a menu of sleep schedules to choose from: the 'bi-phasic siesta' (one daily nap, after lunch), the 'Dymaxion' (Buckminster Fuller's choice – no core sleep, with four half-hour naps per day). The 'Uberman' is popular but unfortunately named; it involves six twenty-minute naps, without a core sleep. The internet is full of accounts written by people following these programmes, very few of whom last long before returning to a traditional, one-sleep schedule.

If you keep yourself awake too long, scientists have found, your body goes into a kind of sleep-starvation mode; once you crash, it will be a more powerful 'recovery sleep' which can restore the body in a shorter amount of time. Polyphasic sleeping sets off the same physical response; the naps grow more and more effective, correlative to how exhausted you are.[10]

Polyphasic sleeping has been used by NASA, the Canadian Marine Pilots, the US military and long-distance solo boating champions as a way to manage sleep in extreme situations, but most of its fans apply it to little more than an extremely intense workday. Its advocates – at least, those I have seen online – are almost always young, and are almost always men, willing to sacrifice their social lives and sometimes their well-being in the name of efficiency.

More than anything, it strikes me that polyphasic sleeping is a way to reclaim time when it slips away from you – to compete, and to gain an edge, by bending nature to the demands of the 24/7 remote workplace. It falls somewhere between biohack and salve for the overworked mind; who even knows if it makes people more productive, but it's a way for them to cope with the feeling of not doing, or being, enough.

'To really give polyphasic sleep a run for its money,' one writer, Danny Flood, states in his 'Ultimate Guide to Polyphasic Sleep', 'you have to have the right things to fill up your waking hours. You have to be busy ... preferably doing epic shit ... Writers, by far, are the best suited for polyphasic sleep because there is always more to write.'[11]

This is a feeling that's all too familiar. What's also familiar is the loneliness that sets in when you stray too far from your own biology; there is a certain community in sleep, which you lose in renouncing it.

These new models for sleep imply a rejection of the body, perhaps even disgust with its needs and frailties. They require you to admit that something is wrong – that you're struggling to keep up with the world, as it currently stands – but instead of going with a solution that feels natural, you must find the most extravagantly difficult option, and pursue it.

Every insomnia narrative is culturally constructed. Sleep is a cultural construct too.

This admittedly rash statement appeals to me, but then, I've not slept for two days and my judgement is fallible. I'm reading *24/7: Late Capitalism and the Ends of Sleep* by Jonathan Crary, one of many books about the politicisation of sleep, which argues that capitalism conspires to keep us from sleeping. I'm also reading research papers, mentioned by Crary, which examine the creation of a 'sleep crisis' in news media. As the hours go by my attention diminishes, and I switch to tabloid news articles about how sleeping pills are causing dementia, are used by suicidal people to overdose, and how a mother of three in South East London once mixed them into a chilli con carne in order to poison a husband who was about to divorce her. I read about the rapper Post Malone, who has 'Always Tired' tattooed across his face, under the eyes where the shadows tend to show. My favourite is a piece in *The Sun* warning of the risks of sleeping pill addiction, with three different women – all of them, it is mentioned, are mothers – photographed wearing pyjamas and smiling, and talking about their secret sleeping pill hell.

I dredge through archives, posts marked 'insomnia' published by the *Daily Mail* and the *Guardian*, and items shared to Reddit and Twitter. Each year, and month, and day brings new and contradictory stories on the state of the sleeping world. Is there really a sleep crisis, or is this a sleep culture instead?

In the final week of my stay in London, I somehow cycle through the entirety of *It's Always Sunny in Philadelphia*, seasons one to twelve, on Netflix. This is the fourth time I've watched it all the way through. I mean 'watched' very loosely; although I love the show, I'm not really watching it, even when I'm awake. Rather, I'm using the show to keep me company, in a slowed, semi-zombified state, my head propped up with pillows and my mind propped up with a 2 a.m. mug of 'L'Or Classique' instant coffee, in the hope that it will eventually make me crash and fall asleep. The familiarity of each line is fitfully amusing. I hear my own laughter echo around the empty studio, and I laugh again, out of misery and bemusement at the fact that I have become someone who sits alone, unsleeping, in a darkened room for weeks on end with only a laptop for company.

At Summit LA, a conference held in November 2017, Reed Hastings, the CEO and co-founder of Netflix, publicly named his company's main competitor as sleep: 'You get a show or a movie you're really dying to watch, and you end up staying up late at night, so we actually compete with sleep ... And we're winning!'[12]

This idea deserves further scrutiny; rather than naming Amazon, or established TV networks as his rival, Hastings decided to take on human biology. Hastings doesn't just want us to stay awake longer to watch Netflix; he's positioning Netflix as a new kind of entertainment, one specifically designed for consumption in the intimacy of your bed. Even a bedroom TV placed on a wall is a little more distant; he's talking about laptops balanced on knees, smartphones held up to faces, screens of all sizes glowing close to the body in the dark. He describes Netflix as a pointedly antisocial activity; you could go out with friends, you could 'relax and unwind, hang out, and connect', but instead you're at home watching TV.

Around 75 per cent of the shows viewed on Netflix are driven by its recommendation algorithm. Netflix also tracks the 'completion rate' of shows, what day and time you watch its content and the device you watch it on, along with your scrolling and browsing behaviour. The 'post-play' function, where a show plays automatically after another ends, is a very unsubtle, very calculated move to keep you hooked, paying on average $0.20 per hour of content, with roughly 1 billion hours viewed every week, or 140 million hours per day. As with all compulsive online activities – social media use, gaming, gambling, porn – there is something distinctly solitary about the Netflix experience, which relies heavily on algorithms and personal tracking to predict what you'll want to watch next. It's not much of a stretch, then, to argue that Netflix is for sad people. For lonely people, at least.

Research indicates that streaming is mostly a solitary activity. Shared accounts are no proof against this – rather, they allow families to watch different shows, alone together in different rooms and on different devices. It also seems that binge-watching, specifically – the kind that Netflix encourages – is linked with poor mental health. A survey of binge-watchers conducted by the Universities of Michigan and Leuven in 2017 linked it with fatigue, poor sleep quality, insomnia, depression, anxiety and feelings of loneliness.[13] This has become a crisis narrative of its own, within the broader 'sleep crisis' argument. A survey conducted by mattress company Amerisleep in 2017 listed reasons provided by those who blamed Netflix for their insomnia; 25 per cent blamed the autoplay feature, 29 per cent said they were 'zoning out and losing track of time', and 46 per cent said it was 'the addictive nature of a favourite series'. There's an article on this subject in *Psychology Today* I find darkly amusing; titled 'Binge Watching and Its Effects on Your Sleep', it advises readers to watch Netflix with someone else; '... in groups, someone is more likely to step up, say, "enough," and turn the TV off. Alone, we're more likely to let binging impulses get the better of us.' If you must watch alone, the piece suggests making an appointment to call a friend afterwards.

Perhaps this is what the phrase 'Netflix and chill' was originally intended to mean, a world where Netflix is both sedative and distraction. An 'Entertainment' in the *Infinite Jest* sense of the word; one so addictive and convenient and blandly, compulsively agreeable that you'll keep watching at a cost to your bodily functions. Ideally you won't sleep at all, but if you do, you will allow Netflix to lead you into sleep, to watch over you for a span of two or three episodes before tactfully leaving and allowing your laptop to switch to a screensaver.

Scientific studies around Netflix and sleep, and media and sleep in general, focus on the connections between watching videos and insomnia. But it's hard to know which came first, whether using these services causes insomnia, or whether insomniacs tend to gravitate to them in the first place. Speaking from experience, when I'm sleep-deprived I'm in no mood to read or write, or to do anything else remotely productive, and I more often end up watching Netflix alone in bed.

Most of the time, when I'm awake at night with Netflix, it's not because I'm watching something addictive and new. I'm more often watching old shows, where I already know the plots and jokes almost by heart. What I'm addicted to is noise, distraction, and a kind of automated company. But my mind, it turns out, is louder than *It's Always Sunny in Philadelphia*.

At night I scroll through social media timelines. I scroll for miles, until my eyes hurt and my laptop's cooling fan is hissing. It's always infinite, a race to the bottom with no end.

Infinite scroll was popularised by Pinterest and Twitter, and later Facebook's introduction of the Timeline in 2011. Platforms love infinite scroll because it keeps their users hooked. It speaks to the brain's endless search for novelty; the thing we want is always hidden, just off the screen, so we keep looking, dopamine spiking, brain held on the brink of receiving a reward.

After a while, I find, the content of the site I'm scrolling through doesn't matter; it's the promise of the unknown, the repetitive motion, that keeps me there browsing without end. We're not even reading as we scroll, not *really*; one study, conducted by University College London back in 2008, concluded 'It is clear that users are not reading online in the traditional sense; indeed there are signs that new forms of "reading" are emerging as users "power browse" horizontally through titles, contents pages and abstracts going for quick wins. It almost seems that they go online to avoid reading in the traditional sense.'[14] One Republican senator, Josh Hawley of Missouri, even proposed a ban on infinite scroll in a bill, intended to save the public from internet addiction (the bill would also have banned autoplay videos and limited users to half an hour of browsing time per tech platform, after which they'd be alerted and asked if they needed more time).[15]

The idea of banning a form of UX design seems extreme to me, and too much like admitting human defeat in the face of technology itself. Yet that's often how I feel when I lose hours of the night, turning sleep time into screen time; I feel futile, as though my emotions, and my critical faculties, have been short-circuited by the internet. I'm not an addict, though, because an addiction is defined by its ability to get in the way of everyday life. The internet isn't in the way of my life; it *is* my life, and I find it very difficult to imagine an alternative.

Three weeks into my stay in London, and deep in insomnia, I take the Tube to Old Street to visit a nap hotel. I do this out of curiosity rather than necessity – I want to experience sleep as a product, to visit the crossroads where sleep and capitalism meet.

On its website, the nap hotel is marketed to business travellers, 'to improve their well-being and mindfulness while being away from home'. I've read about similar places in Japan, which usually involve some kind of pod or tiny, futuristic sleeping cubicle. They're aimed at salarymen who don't have enough time to go home – men who have long been the target market for energy drinks like Lipovitan-D, and who have unintentionally adopted a polyphasic sleep schedule, out of necessity rather than as a lifestyle experiment.

Perhaps the nap hotel is meant as a status symbol; only the most important, most sleep-deprived people could possibly be busy enough to require it. I book a time slot and pay ahead with PayPal. It costs £8 per half hour – I opt for an hour – and is located between Old Street and Shoreditch, just off Silicon Roundabout in the area known as London's 'Tech City'. Getting there is complicated; I worry that I'll be late for my nap. I take the Northern Line to Old Street, but once I arrive I fall into one of those curious cycles of Google Mapping, where you zoom in as much as possible on the screen but keep moving in circles.

Insomnia lends the world around me an air of the fantastical. It's hard to break Old Street Roundabout into individual scenes and spaces, because it exists as a whirling whole. There are pop-up shops, and marketing agencies specialising in pop-up shops. There's a co-working space, one of many, called the White Collar Factory. There are also posters for apps catering to the shut-in economy, which deliver food and dry cleaning, necessities like razors and artisanal coffee, so that customers can stay within their cycle of work and home without ever needing to go shopping.

Like shopping malls, and like Dublin's tech district, this area has its own psychology, a series of statements made, in the form of oddly shaped, expensive buildings, which add up to an indistinct grey blur when experienced from the ground. Old Street's momentum compels me to keep moving, glancing down occasionally at my phone for directions. I pass a street called Silicon Way and a series of hotels, and then I encounter the most alarming building I think I've ever seen.

The M by Montcalm Shoreditch Tech City Hotel, on City Road, is a luxury hotel built in 2015. It's pale grey and jagged, dotted unevenly with windows shaped like vertical slits. There are eighteen storeys, bizarrely slanted so that the building resembles a sinking ship, as though inside each room the angles tilt like the set of *The Cabinet of Dr Caligari*. The internet tells me that at the M by Montcalm, the receptionists take your details on iPads, and that tablets are built into the surfaces of tables in the lobby. There is the option of a 'Scent Steward' for your room, who will help you to select what it smells like, as well as a 'Bath Butler', who chooses bubble bath, and a 'Sleep Concierge', who helps you choose from a menu of different pillows.

As I draw closer, the building only becomes more confusing; the roof rises to a point in scalene horror like an optical illusion, shanking the sky. The sight of it triggers a mental transposition in me, violent and sudden, prompting visions of a hell populated with skyscrapers. The 'calm' in its name strikes me as especially hilarious. If buildings had soundtracks, the M by Montcalm's would be the violins in *The Shining*, or maybe the sound of bees used to signal the presence of Pazuzu in *The Exorcist*.

The M by Montcalm is marketed as an aspirational place to sleep, but looking at it from outside I feel only horror. I walk past, turning deliberately away. The nap hotel is down a little road between towers. I pass two large buildings occupied by WeWork, and remember reading that membership, which costs upwards of £400 per month, includes 24/7 access. Slogans in gigantic, calligraphic script decorate the door; they read 'Do What You Love'.

Finally I find the nap hotel, down a lane and up a set of stairs. The company is billed as a 'premier wellness start-up'; along with running the nap hotel they work with corporations, setting up 'sleep pods' in offices and giving talks on the importance of sleep.

It seems no coincidence to me that the nap hotel is located in London's tech district. It appeals to a client well-versed in sleep hacks as well as sleep crisis, whose work has disrupted their sleep, but who now has enough money to pay for naps. A business like this caters to the sleep philosophies of both Arianna Huffington *and* Elon Musk; by paying for naps you are prioritising your sleep, but also making a point of being sleep-deprived enough to require their services in the first place. The nap hotel is the public face of a corporate wellness strategy, one where a healthy sleep is 'taught', and encouraged, in the name of earning money.

Inside the nap hotel, the air is perfumed with lavender and vetiver. The light is dim. It's utterly silent, as silent as possible when you're several minutes away from Shoreditch and Old Street Roundabout. The walls are devoid of colour – black, white, and a deep shade of grey – and the interior has been segmented into private cubicles, each one occupied by a bed and a bedside table. The nap hotel's solitary staff member is friendly, but seems surprised to see me. I suspect very few people actually come here, and wonder what he does for most of the day, spending his shifts in lavender-scented tranquillity. He leads me to my 'pod', as it's referred to on the website. It's more like a monk's cell, or a room used for massages. There's a single bed, a bedspread that same grey shade, a potted plant in the corner, and another on a bedside table next to a lamp, an eye mask, a glass of water, a marketing flyer made of white card, and a small wooden artist's model of the human body. The walls are not walls, but heavy curtains. This doesn't do much for privacy, but it does preserve a certain calm, a monastic atmosphere of mutuality. He pulls the curtain closed and silence falls.

I do not sleep in the nap hotel. Instead I type out six emails on my phone and send them to myself. They comprise my notes, for writing this essay.

Alone in the booth my movements become careful and slow, pulling my shoes off without bothering to untie the laces, picking up my phone and putting it down again. I worry that my stomach is making lurching sounds; I haven't eaten since breakfast. My phone vibrates slightly every time I touch one of the keys. I turn this function off, as I suddenly feel very worried that the receptionist will hear it and think I've brought sex toys to the nap hotel.

Almost two weeks ago I started taking 5-HTP, a supplement I bought from a health food shop at Piccadilly Circus, and I suspect that this is adding to the experience. A friend told me it gave him interesting dreams. I've not had many dreams, personally – possibly I haven't slept enough – but I have found that it helps me feel happy, so I keep taking it. A few hours after swallowing each pill there's a sense of 'coming up', a buzzing feeling of energy and purpose, and this coincides, perhaps unfortunately, with my visit to the nap hotel. Lying in the dark-grey bed, behind the curtains, I feel an overwhelming, almost certainly chemical intensity, as though my brain is speeding forward but going nowhere.

There's a pressure in the front of my head, a pleasant yet acute sensation. The scent in the air; the calm; it all makes me want to cry somehow. It bears a tiny hint of the cosmic yawn induced by MDMA, or how I feel on benzodiazepines after roughly an hour, like my brain cells are stretching. I stretch out my legs on the bed, and my bones make a clicking noise, loud enough that the receptionist can probably hear. Fuck it. He has probably heard stranger things, in the course of working here.

The eye mask on my bedside table looks like a black satin bra, made for someone with an unusually prominent sternum. Moving slowly, silently, I pick it up and place it over my eyes. My thoughts are not quieted. Minutes tick by. I lift the mask to glance at my phone; it's almost a quarter past four. I wonder if coming here was a bad idea. Forty-five minutes suddenly feels like a very long time.

The nap hotel is a reminder that silence is considered a requisite for sleep. I've not observed that for years; in my life, the border between noise and silence has collapsed, just like the borders between night and day, working and off-time, anxiety and leisure.

Four forty-five at the nap hotel. I briefly try to meditate, and last less than two minutes before checking my emails again. I notice that the room is cool; I've read that the best temperature for bedrooms is between 16 and 18 degrees centigrade, as it helps your body lower its temperature for sleeping. The vetiver and lavender smell is still present. I lean over and sniff the plant on the bedside table, in case this is where the scent is coming from, but it turns out to be made of plastic. Next I investigate the mattress. It's thicker than I expect, and beautifully consistent – spongy and thick, but not too firm.

At home in Dublin, I sleep on a terrible mattress. It's around three decades old and viciously lumpy, sagging inwards at its sides and spilling off the edges of my old wooden bed. I have slept on it for so long now that I have memorised a topography of its lumps. The lumps are a second presence in the bed; sometimes I wake in the night and know to re-adjust myself around them. At the nap hotel, the mattress is infinitely more comfortable. It would be useful as a mattress-sampling service, that is, if I had plans to buy a new one.

It's 16.55 now and I've given up on even pretending that I'm sleeping. Place, it seems, cannot force sleep upon a person, nor can sleep respect a place designed for its purpose. The nap hotel is listed on Google Maps as a 'Wellness Programme', and in a sense I suppose that's what it is. It's a fleeting, calming experience, perhaps less calming *because* it is fleeting. It's an experience costing 26p per minute, engineered to make you feel calm enough to forget about the cost.

The nap hotel reminds me of the value of sleep, and the value of silence. Such is the sound of capitalism – car horns, planes, trains, commerce and noise – that even silence exists on its terms, and at a cost. If a job assigns value to time spent working, then the nap hotel is its inverse, and its dependent, because now time spent sleeping has value too.

Would I be a more productive citizen if I slept more? If I drank less coffee? If I did yoga? My therapist certainly seems to think so. But isn't it also in the interests of work that I pollute my body, exhaust it, wring from every day the upper limits of its potential?

Years ago, during an argument, a very techy, very profit-minded friend of mine told me that his time was money, and that I was wasting his time. We fell out for a few years after that. But my time is money too; I don't want it to be that way, but it is. It's just that I'm more often spending than earning, and that my earnings are segmented into haphazard blocks. The nap hotel, selling naps by the half-hour, raises the jarring but pertinent question of how much sleep costs, per second, per minute, per hour, per life.

*

It's dark outside when I leave the nap hotel. I happen to have caught the precise hour, in London in winter, when the grey dims down further into night. On the Tube I glance at a leaflet I took from the bedside table in the nap hotel, which reads 'Sweet dreams start here: The ultimate power nap ... on five layers of comfort.' Inside there is a diagram of a mattress, the same mattress I have just spent an hour resting on. It's made from 'natural cooling soybean balms' and Visco memory foam. The flyer, it turns out, is a voucher, offering me £100 off the next mattress I buy.

I'll probably never buy a new mattress. I don't have the money, and besides, I don't sleep, not even when I've paid to visit a location designed specifically for taking naps.

Today Edison's quest against sleep feels like an augury; his attitudes have filtered into everyday life. Technology intrudes on sleep, and drafts it into an agenda of profit, surveillance and competition. Sleep has become another thing to disrupt. No wonder, then, that I can't sleep, with so many voices telling me *how* to sleep at the same time.

We've come a long way since the days of the Psycho-phone whispering mantras in your ear. Now I scroll for knowledge through the night; one blog post I find, written by data scientist Leo Qin, estimates that the average internet user consumes 6,103 lines of text per day, and scrolls through 74.21 feet on screen daily. He calculates an average scrolling distance of 5.03 miles per year.[16] I've been online since I was twelve years old, or thereabouts; in total I may well have scrolled 90.54 miles, roughly the distance from Dublin to Roscommon, or halfway across Ireland.

Perhaps there are ways to reconfigure sleep that fall outside the orbit of present-day sleep-hacking. A historian, A. Roger Ekirch, is known for arguing that before the Industrial Revolution people slept in bouts, following an early version of a polyphasic schedule. The night was divided into a 'first sleep' and later a 'morning sleep', with time for reading, writing, sex, visits to neighbours or acts of petty criminality in between.[17] Debate continues as to whether Ekirch is correct, and if he is, what it would mean for our culture's current sleeping habits.

Another historical treatment of sleep is that night was always reserved for sleeping, because sleep was 'an honest man's rest'. Sleep was embedded deeply in popular morality, mainly for the purpose of creating energetic workers. *Narcocapitalism*, by Laurent de Sutter, explores this morality of night. De Sutter writes, 'For all those who looked favourably on the development of industrial capitalism, this assessment was something of a principle: what was needed were individuals who slept soundly – so that they could get on with their work unhampered the next day.'[18]

This same issue plays out today in corporate talks given by the people who run the nap hotel, and in articles published to Arianna Huffington's Thrive Global; burnout is loss-making, and sleep is essential for productivity. I see far less discussion of the nature of work in our current technological age: diffuse, precarious, and no longer confined to eight hours, or even to a traditional workplace. Work is anywhere, and any time you're able to connect to the internet. An 'honest man', then, must rearrange his sleep schedule around his work; he should probably stop sleeping.

The more I look around the internet, the more I realise it's full of solitary insomniacs; people on YouTube, alone, talking to cameras. Sadposters; depression-posters; people who post, then delete their late-night tweets the following morning. People paid to work at night. People on one side of the planet, providing customer services to people on the other. People who, like me, go to the gym at night, or who channel their premenstrual crying jags into scrolling through pictures of dogs, collected in dogspotting Facebook groups by other emotionally fragile people.

Those nights of insomnia left me feeling like I was always running behind – behind on work, on contact with friends, and behind on daylight. For all the extra hours, I was rarely productive. Doing nothing never felt like such hard work.

Loneliness comes naturally to me, and night brings me closer to my computer. This essay was written in the early hours, across a series of very long nights in December 2018. I will finish it as I began: alone, wearing pyjamas, with too many tabs open on the screen and a cup of coffee going cold by my side.

While writing, my eyes have begun to hurt. They've taken on a strange, grey-red shade, appearing half-open no matter how much mascara I apply. The skin under my eyes is hollow, and tinged with purple and blue. I don't look like myself; I'm struggling to recognise my own reflection.

Perhaps it's time to stop now, and finally sleep, the screen still glowing in the dark beside me.

All Watched Over 2: The Best Sleep

JUST BEFORE CHRISTMAS I LEAVE LONDON, where I have been sleepless for most of the month, and go back to Dublin. I'm finally observing a more normal sleep schedule. Sleep came back slowly; the heaviness, and the half-open eyes. I began to surrender, taking naps during the day, then late in the day, then easing myself into sleeping at night and waking in the morning.

The more I thought about sleep, the less I was capable of sleeping. Yet the more I read about it online, the more I felt like I wasn't thinking about sleep enough. Modernity, it strikes me, doesn't leave sleep to chance. Not when its alternative – the controlled, perfected sleep, a mirror of the productive work day – is so difficult to attain, and so lucrative.

The first story published by William Gibson, in 1977, is called 'Fragments of a Hologram Rose'. In it, a man called Parker has trouble sleeping and relies on a technology – 'ASP', which stands for 'apparent sensory perception' – to lull him to sleep. It shows him footage shot from the perspective of a 'young blonde yogi', who does exercises on a beach in Barbados before falling asleep. The device blurs the subject's perception with that of the viewer, inducing calm.

ASP could be Gibson's premonition of VR, but it might just as easily be like ASMR, where video and audio are allowed to intensely manipulate the senses. Personal memory and film intermesh; the ASP carries Parker back to a series of scenes from his own life, where his girlfriend leaves him and takes a rickshaw into the anonymous night. Parker is world-weary, haunted and tired, and the device manipulates his emotions. We're informed that he hasn't been able to sleep without an 'inducer' for two years; he's addicted, a cyborg whose prosthesis is sleep.

Gibson is known as the pioneer of cyberpunk – science fiction set in a dystopian future, characterised by 'low life and high tech'. But his portrait of technology as a cure for insomnia and loneliness might not be far from our present day.

Today, for every lonely person kept awake by notifications and the endless scroll, there's someone else sleeping *with* technology, pursuing an augmented, monitored sleep. Products are marketed to those seeking to achieve it, and communities have evolved around sharing experiences and experiments in sleep-hacking.

Sleep technology has progressed from science fiction to reality, and the internet is a battleground for companies seeking to turn our eight hours into profit. The most accessible, commercial face of this trend is the current market for mattress start-ups; it's likely you've noticed ads for them, sponsoring podcasts, or paying to be mentioned by Instagram influencers. They're companies based entirely online, rather than in showrooms, which deliver mattresses they claim are engineered with varying degrees of innovation for the perfect night's sleep.

After my visit to the nap hotel I looked up Simba, the company that had provided it with mattresses, and fell down a rabbit hole reading about the 'mattress wars'. Casper is likely the best known of these brands, with Eve, Emma and Otty, among others, biting at its heels. They're called 'bed in a box' companies; once your mattress is delivered, you transport the box to your bedroom, unfurl the mattress, and it puffs up to its full, decompressed size.

These start-ups emphasise technology and innovation, framing sleep as a ritual of competitive self-betterment. In search results, Casper is 'an obsessively engineered mattress at a shockingly fair price'. Leesa has a 'universal adaptive feel' and 'combines cooling Avena foam with pressure-relieving memory foam'. Ghostbed has 'supernatural comfort with cooling technology', while Purple is 'the world's first comfort tech company backed by science'.

All of the above are excessively, exhaustively reviewed on dedicated mattress sites and in YouTube videos. The reviewers, by and large, are young, wholesome-looking men; they manage to take all implication of sex and even intimacy out of bed-reviewing. Their tone is earnest. They wear freshly laundered hoodies and occasionally mention their girlfriends and wives. They also profit from affiliate links; with each click-through from a potential customer, the blogger takes a small cut of their purchase. In videos, we watch them sleeping, or at least pretending to sleep. The internalised surveillance of social media becomes even more pronounced here; 'sleep marketing' falls at an intersection between consumerism, biopolitics and the quantified self.

What most of the reviewers I encountered had in common was their use of the phrase 'the best sleep'; the sleep all their efforts are in aid of, which is achieved not through relaxation, but through consumerism and intensive self-monitoring. The reviewers wear devices to track their hours of sleep, reporting on whether the mattress was good for back pain or side-sleeping. The mystery of sleep can be harnessed and controlled. The best sleep is a consumer quest; the best sleep is something that you can buy.

That the 'bed in a box' companies target young people as well as older, more settled consumers is significant; it implies that young people, less likely to own property or even a car, and destined to move between a series of rented apartments, would however be willing to take a mattress with them between these moves. The rising number of divorces and the number of millennials waiting longer to get married has been kind to the mattress industry; the longer we stay alone, the more individual mattresses we're going to need. Your mattress is real estate for life (or, rather, for five to ten years, after which you're encouraged to replace it). It is your home within a home; your escape; your guardian angel; the expensive, foam-padded keeper of your body and soul.

Technology steps in to monitor and control sleep, even when the brain has no choice but to switch off. Once you've found the perfect mattress, you can enhance your sleep further with apps and wearables. Sleep tech is a popular theme on Kickstarter, where crowdfunded projects include an adjustable pillow with a corresponding phone app, a wifi-connected alarm clock which doubles as a 'sleep revolution device', the 'Bedjet 3' climate-cooling air-conditioning system for beds, a 'Sleep Shepherd' brainwave-monitoring nightcap and the 'Gravity' weighted blanket, which was advertised to me on Facebook. Meanwhile in more mainstream technology, in 2016 Apple introduced the 'Bedtime' feature for the iPhone, a sleep-tracking app that can be configured to send alerts reminding the user it's time to go to bed.

I have experimented with sleep-hacking over the years, less out of curiosity than from desperation. I've taken a consumerist approach to sleep, seeking out pills to make it easier. I've fallen into online communities, reading their stories and comparing their experiences with my own. Time passes, and then my insomnia comes back. Perhaps I'm not trying hard enough; perhaps I need to lean into my own mutation.

The mattresses, the apps, the nootropic supplements – all of them promise to deliver 'the best sleep', the sleep our bodies long for but our minds have forgotten. It comes at a cost; in darkness, as in daylight, technology estranges us from our senses.

Nature recedes, in memory, on the screen, until it exists only in dreams.

Drug: Zopiclone[1]
Exp. year: 2012
Dose: 7.5mg
This was my first sleeping pill, and I was surprised by how effective zopiclone could be. Suddenly it felt like sleep could be controlled; I could switch myself off, like a computer. I didn't dream, and when I woke I tasted metal, like the inside of a robot's mouth.

Back in 2012, when I lived in London and worked on social media as a cheese, I began to believe that I was stronger than sleep. My work, and the city, encouraged it; my job was as an 'always-on' marketing creative, running social media accounts during the day, but also checking on them in the evenings and over weekends. After work I tried to get out and see people, but always came home exhausted. The city was a challenge; earn enough to live there, and live enough to make working there worthwhile.

That was the year I had my first proper bout of insomnia. I'd never been much of a sleeper in the first place – apparently I hardly slept as a baby, and this continued into childhood, when I would read past my bedtime, late into the night. In adulthood this meant roughly six hours per night, but in London this dwindled to three or four. I couldn't stay asleep; I started waking at 5 a.m. every morning, and never knew quite what to do without making too much noise, and waking up my housemates.

Falling asleep, too, had become a problem; I would lie in bed, sometimes in silence, sometimes listening to podcasts. Hours would pass, and behind my eyelids a cruel automated slideshow would replay past failures and humiliations, stupid things I had said, break-ups, badly worded things I had written. My brain had turned against me, beyond my control. Falling asleep, and staying asleep, suddenly became a task requiring skill. I was failing at it. It was only in losing control over sleep that I realised I'd had control over it in the first place.

I was able to see a doctor through the NHS. After several minutes browsing through a gigantic book – a catalogue, I assumed, of psychotropic medicines – she asked if I'd like to try zopiclone. I didn't have much knowledge of sleeping pills, but I found the prospect interesting, and said yes. Zopiclone is a hypnotic pill used to treat insomnia, one of the 'Z-drugs' (the 'Z', I like to think, is for 'zombie'). It's classed as a cyclopyrrolone, similar to benzodiazepines but functioning differently. Brand names include Imovane, Dopareel and the pills I was prescribed, Zimovane.

Collecting my prescription, I experienced the same vague sense of excitement I always feel when I try a new pill, akin to buying new shoes, or getting a haircut – a hope that this might potentially be a 'new me', a way to patch up my flaws and start over. It would help me blend in with the world and stick to a schedule, showing up at my desk at 9 a.m. each morning. I'd already taken time to adjust to the demands London placed on me – just getting around the city required more energy than any other place I'd lived. The pills were good for knocking myself out; I became familiar with how long they took to kick in, roughly half an hour before I'd pass out as soon as I was in bed. They didn't keep me asleep – I continued to wake up at five every morning – but it was enough to get by.

Upon waking I began to notice evidence, left on my computer, of my actions in the minutes before I'd fallen asleep. There were long, rambling chats on Facebook, Gchat and Skype with people I didn't know very well. Taking zopiclone was like becoming suddenly, violently drunk; it bred a casualness of tone, an assumption of easy intimacy with strangers. If I didn't resign myself to going to sleep by the time that they kicked in, I would chat with people on OkCupid, a dating app, where my preferences were suddenly different, and stranger. Under its influence, I would Like profiles of people I'd never usually go for, and the conversations I had with them contained a queasy mixture of clumsy sexuality and unhinged humour.

On the nights when I took zopiclone I was myself, but re-arranged in an unfamiliar way. I was a less consistent, jaggedly paced version of myself, like I was dreaming while I was awake. I acted against my personality; I ate all the food in the fridge, and compulsively cleaned the flat, poorly. One morning I woke with the usual zopiclone-induced metallic taste in my mouth, and beside me was a broom, tucked under the blanket with the handle poking out. I found a neat little pile of dust in the corner of the doorframe. I had swept my bedroom, the kitchen and the hall.

As it turned out, all of the above is relatively normal behaviour for someone on sleeping pills. Searching for information online I found communities dedicated to discussing them; on r/ambien, the subreddit dedicated to Ambien, zopiclone's pharmacological cousin, the pills occupy a space between psychedelic high and medical obligation. Sometimes there are posts looking for advice on dosage, often for first-time users, while others discuss tapering off after long periods of Ambien usage. There are numerous posts written while under its influence, 'trip reports' about combining Ambien with other drugs, and occasionally users asking, 'Am I going to die?' Revisiting r/ambien today, I can't help but laugh at some of the comments: a nonsense post headlined 'AAAAAAA-ZOR , 7guy ok' is followed with a comment from the OP, saying, 'I woke up to this post and I really don't know how to feel about it.' Someone else has posted pictures from three different visits to Disneyland on Ambien; in the most recent, he's wearing Minnie Mouse ears and an Aladdin T-shirt, and gnawing dazedly on a gigantic leg of turkey, his eyes half-open and fixed in consumerist ecstasy.

It was on the Ambien subreddit that I first encountered the Ambien Walrus. Comments mentioned how 'the walrus made me do it', or that the author had woken up to the consequences of its whims, memories lost in a pharmaceutical blur. The Ambien Walrus first appeared in a webcomic called *Toothpaste for Dinner*, in 2007, and was shared on blogs before spreading to Reddit. Today he's the unofficial mascot of sleeping pill users, a hybrid of demon and guardian angel. A little like the Coke Badger from the film *It's All Gone Pete Tong*, the Ambien Walrus manifests right around the time your sleeping pills kick in, then keeps you awake offering dubious advice. The human, in these cartoons, is always crudely drawn, and appears fragile and terror-stricken. 'Oh god,' he says in a *Toothpaste for Dinner* comic of 2009, 'the Ambien Walrus ... I must have forgotten to go to sleep.' The walrus responds in all-caps: 'TAKE SOME MORE AMBIEN AND CUT OFF ALL YOUR HAIR MAN LET'S DO THIS.'[2]

What appealed to me most of all about the Ambien Walrus comics, and about the Ambien community in general, was how it brought together uniquely vulnerable people. Insomnia is a lonely state, and taking a pill for it means venturing into the unknown, trusting your unconscious body to science. The Ambien subreddit is a place for insomniacs to share their experiences, to pool knowledge, and to document bizarre, uncharacteristic behaviour which the pills wipe from their memory the next day.

Sleeping pills are an imprecise technology, one where the flaws are still being ironed out, and where you choose between remembering (often too much), and remaining awake, or sleeping but also forgetting (for this reason, zopiclone also reminds me of the 'Forget Me Now' pills used by 'Gob' Bluth in *Arrested Development*, which trap the hapless character in a 'roofie circle' of amnesia and addiction). To compensate for one faulty technology, its users turned to another; Reddit's r/ambien, anonymous as it is, offers a refuge for people waiting for their pills to kick in, seeking virtual company and reassurance, but who cannot risk speaking publicly in their intoxicated state.

With zopiclone the effects were almost instantly tangible; I had managed to hack sleep. It felt like I was in on a secret, in league with a twilight society of pharmaceutical dreamers. But eventually I ran out of zopiclone; I had my prescription renewed once, but not a second time, and suddenly I was back where I'd started, forced to confront the issues which had disturbed my sleep in the first place, like hating my job, and being depressed, and needing to make changes to my life in general.

To me, zopiclone suggests that the 'best sleep' might be a medical myth – it's a drug that achieves sleep, but with side effects that make its use untenable. I think I might have caught the end of a wave of zopiclone usage, right before doctors began to view it as addictive, and started prescribing it less frequently. In 2014 it was made a Class C controlled drug in the UK, with penalties for anyone caught selling it. Researchers noted an increased mortality rate associated with Z-drugs, and news reports named it as hazardous, causing old people to spontaneously fall down, creating amnesiacs and dangerous drivers (it was found to increase the risk of next-day vehicle accidents by 50 per cent), and, when combined with opiates or alcohol, causing coma, depressed respiratory function and death. Zopiclone was used in overdoses by suicidal people. It made people spontaneously violent, prone to 'sleep sex' and hallucinations. You could get high on it by exceeding the recommended dosage, or combine it with alcohol to induce a state of euphoria.

When I moved back to Dublin, I read reports about the streets around the North Strand, where I later ended up living. Drug dealers had taken to selling 'zimmos' – Zimovane – during a heroin drought. Tablets were smuggled through Europe from

China and Pakistan, and sold on Irish streets for roughly €10 for a pack of fourteen pills. Its consumers were often people suffering from years of heroin addiction, who without Z-drugs could not sleep from the pain of withdrawal. A reporter for the *Irish Times* interviewed a group of people taking Zimovane in daylight. One said, 'Zimmos make the day go by without you noticing it.'[3]

Drug: Sertraline
Exp. year: 2017
Dose: 50mg
The Sertraline prescription marked my fourth attempt at treating depression with an SSRI. One of the strange side effects was suppression of REM sleep, meaning I woke up more, and was more likely to remember my dreams. In the daytime, my hands shook constantly. I experienced brain zaps, headaches, intestinal pain and little evidence of ever feeling happier. But strangest of all was my new-found tendency to wake up screaming in the night.

Much of the anxiety I experience at night comes from social media rather than real life. The internet has primed my mind for distraction, to the point where sleep seems distant, and impossible.

Before Netflix, I used to be a little more high-minded; I'd fall asleep listening to podcasts about economics (I'm not sure I really took in any of their content, but at least my intentions were lofty), and *This American Life*. One time its host, Ira Glass, came to speak at the debating society at my college, during my postgraduate degree. I queued to meet him, and told him I thought I'd subliminally absorbed some of his accent. He agreed; my voice sounded oddly mid-Atlantic.

There was also a lesser-known podcast I subscribed to, which had been created especially for sleep. *Sleep With Me* is presented by 'Dearest Scooter', the online alias of Drew Ackerman. Beginning in October 2013, Ackerman has put out a weekly show under the tagline 'The podcast that puts you to sleep'. Each episode begins with an opening ramble, before Ackerman launches into a meandering, quasi-psychedelic story designed to distract and bore the listener. Sometimes these stories involve a fictional reporter called Claude Neon, a superhero called 'Superdull', or rambling, deliberately pedantic discussion of TV shows (in 2014, *Sleep With Me* evolved a spin-off called *Game of Drones*, dedicated to *Game of Thrones* recaps told in Ackerman's digressive style).

I interviewed Ackerman over email in 2015. Rather than typing out answers to my questions, he responded with a series of short audio recordings. Ackerman was as eccentric, as charming and sincere as I hoped he'd be, acknowledging the often overlooked connection between podcast creators and their listeners. 'There's this strange intimacy when you let someone talk right into your ears,' he said.[4] Discussing the nocturnal community that has evolved around his show, he questioned arguments about technology and isolation, but also acknowledged that they're likely true. 'It's this sad but wonderful thing about podcasts. Maybe it shows up some giant ill in our society that we all feel so lonely and isolated.' A degree of trust goes into choosing what we fall asleep to; Ackerman said, 'I think it's almost a brave thing for my listeners to do, to decide they're going to let this guy just talk them to sleep. It's a strange vulnerability. And on the other side of that, I have to be vulnerable too. It's done in a genuine way. I don't think it could work if it wasn't.'

Since that interview, *Sleep With Me* has been featured in the *New Yorker* and *The New York Times*, and its audience has grown to over 2 million monthly downloads. What I find most interesting about the show is that it strays deliberately, flamboyantly away from what we expect from 'sleep media', a hardly conventional entertainment genre in the first place, but one glutted with whale sounds, New Age music and pseudoscientific 'sleep learning' tracks, the descendants of the Psycho-phone. It also moves against the tide of popular podcasts which present themselves as subject-specific, weighty and educational. It's something you can drift off to without feeling guilty, and without the fear of missing out.

When I interviewed him, Ackerman described insomnia as something 'tribal'; he'd suffered with it himself, from as far back as childhood. Lacking any background in sleep science, or podcasting, he created *Sleep With Me* out of intuition, and it worked.

One person's 'best sleep' is another's nightmare; there are those who require absolute silence, while others find it easier with background noise. *Sleep With Me* offers a counter-narrative to that of popular sleep tech; it's simple and accessible, a voice offering a remote kind of friendship. Simply by existing, it makes its listeners feel less alone.

It strikes me that in the coming years, science will further explore this relationship between technology, sleep and loneliness, and perhaps find ways to make it work in our favour, beyond the usual warnings against blue light-induced insomnia and late-night Twitter addiction. Perhaps this will mark a fall from grace, proof that modernity has altered us beyond hope, and that silence is no longer there for us to return to; sleep will no longer act as a refuge from the noise of everyday life. Perhaps this will simply be a step along a lengthy and inevitable track, one where our minds, our bodies and our behaviours develop in harmony with machines.

Drug: Ashwagandha
Exp. year: 2017
Dose: 1 teaspoon (1 to 3 grams)
I was going through a phase where I no longer trusted medical

science, so instead I bought an unregulated Ayurvedic adaptogen powder for €5.95 from a health food shop. Adaptogens are supposed to make you more resistant to stress. This one also has a calming effect; its botanical name is *Withania somnifera*, the latter part of which means 'sleep-inducing'. Unexpectedly, it worked – though not enough to stop me from going back to Valium several weeks later.

I'm waiting for Netflix to create a category of films and TV shows especially for sleeping: predictable programming with a low volume and gently meandering plots, and a cast of softly droning, unthreatening, good-looking actors. Until then, there are the hypnosis videos on YouTube, which I turned to in the past when *Sleep With Me* wasn't soporific enough.

Across 400 videos, dating from 2011, Jody Whiteley has populated YouTube with hours of monologues read in the same steady, monotonous voice. Her subscribers number over 140,000, and her videos have tens of millions of views. Their titles are functional, and redolent of spam – '2 Hours Awesome Rainy Night Hypnotic Bedtime Story', 'WOW This Hypnotic Bedtime Story Real' – but the descriptions below each video are unexpectedly moving:

> May you fall asleep feeling warm, loved, and comfortable, enjoy a long, sound sleep, and awake tomorrow refreshed, energised, happy, relaxed, and ready to do what needs to get done.

Whiteley's videos average between one and two hours long. They begin with the instruction to find a comfortable, quiet place to lie down and listen, and that, should you need at any time to break the trance, you can bring a stop to the hypnosis with a single, sharp intake of breath. Then Whiteley tells you to listen to 'the sound of my voice', a phrase which appears again and again in her videos, as your limbs begin to get heavier, your breath slower, and as you are shepherded into internet-induced sleep.

Something I find interesting about Jody Whiteley is how closely her voice resembles that of a robot. It's deliberate, somewhere between a hypnotherapist's slow pace and the synthetic clarity of HAL 9000 (who was played by Douglas Rain, a Shakespearean stage actor versed in iambic pentameter). Whiteley manages to remove the personality from her voice while preserving its human kindness. She is neutral, as far removed as possible from YouTube at large and its shrill, personality-based attention economy. The trust her listeners place in her is demonstrated in the positive comments left below her videos, where viewers detail their mental health issues and chronic pain, and years spent suffering from insomnia, and thank Whiteley for providing temporary relief. One reports, 'When I woke up today, the whole word was brand new. I spent all day living in the world instead of my head for the first time in months.'

As with *Sleep With Me*, Whiteley's videos are doubly effective for their listeners; firstly in their calming effect, and secondly in recognising the plight of their listeners. Insomnia, and depression too, involve carrying a solitary burden; the comments below her videos are a place to share that pain, and to feel less alone (it's also worth noting that, like Reddit, YouTube does not require real names of its users, which likely helps Whiteley's followers to express themselves honestly).

Whiteley's videos aren't far removed from the *Sleep With Me* podcast, although they rely on focus, rather than distraction, to lull their listeners to sleep. They occupy a benign, New Age sector within YouTube's hypnosis community. Delve deeper and you'll quickly find how hypnosis is sexualised and used to explore themes of gender politics and control, with videos promising 'hands-free orgasms' and fetishistic 'bimbofication'. There are as many people looking to be hypnotised as there are hypnotists

– there's even a service, Hypnochat, which matches you in conversation according to self-declared role and gender. Reddit hosts r/HypnoHookup, r/HypnoHentai and r/GirlsControlled, a surreal spin on the idea of female celebrities being 'Illuminati puppets'. The way people talk in these communities about being put under hypnosis is especially interesting; they crave the luxury of mindlessness, the ability to switch off their brains and their emotions and follow lines of command like a computer, described inventively as 'brain outsourcing'. While they want to be controlled by someone typing, the internet and the distance it affords is part of the fantasy; they want to become like machines, while handing over their agency to machines in turn. Even the hypnotists, in these scenarios, are likely following a pre-written script.

In online hypnosis the medium becomes the cure, the love object, and the fetish. The internet, and, more broadly speaking, the device, become a source of consensual domination. Videos and audio tracks take away the consumer's autonomy; computers induce orgasms and seize control of minds, inspiring slavish devotion and even love, and bringing to life a dystopian dream of sex with the Singularity. Perhaps it's not surprising that people would seek out these experiences in technology, as we already know that it's capable of mind control; we go to bed with our devices every night.

What is the logical endpoint of sleep with technology? Would it mean handing over control, or finding a way to make technology work for us? The hypnosis videos overlap with another kind of video I often watch, ASMR. I've never understood the appeal of the videos that include only 'triggers', the isolated sounds of tapping and scratching. I prefer videos framed as a dialogue with the viewer, with the person in them paying 'attention'. Some videos are even titled things like 'Personal Attention ASMR' or 'Caring Friend ASMR'; their presenters, who are usually female, are charming, attractive and effusively sweet. More often than not they're talking about something meaningless, and I find myself fighting off yawns, struggling to keep up with their rambling, one-sided conversation.

There's something eerie about ASMR; I know it's pre-recorded, I know it's a video, yet somehow technology replicates the feeling of being cared about. They deliver the sensation they promise – the tingling, the calm – and frequently help me to fall asleep. Perhaps this tenuous, synthetic comfort is enough to trick the body and mind into relaxing.

As with hypnosis videos, the demand for ASMR signals a need to commune with our computers. These videos are calming, yet inherently sad; they mean we're turning to the internet for human comfort, and pre-recorded company, though we sleep alone.

Drug: Cocaine
Exp. year: 2018
Dose: Half a bag? I lost track. It was a party, with drag queens.
So much for nature; when I'm on drugs, I feel like I've become my true cyborg self. Staying up all night becomes a good thing, because I'm with other people. Finally, it wears off, and I realise that cocaine is an ugly drug. Getting to sleep on it is a bit like falling asleep on a Ryanair flight; you might not think you can do it. At some point, you might even convince yourself you're losing your mind, or are about to die. But you'll get there eventually. It might just take hours of remaining still, in one spot, with ringing ears and a feeling like an invisible hand tightly clutching the top of your head while you contemplate oblivion.

'San Junipero', the Emmy-awarded fourth episode of the third season of *Black Mirror*, features a digital afterlife where the dying and the almost dead can relive their youth in their era of choice. The titular beach resort, San Junipero, is a 'party town' where Yorkie and Kelly meet, fall in love, and negotiate the ethical quandaries of transhumanism.

We meet Yorkie and Kelly as young women at a disco in an era that looks like the 1980s, but outside the virtual world of San Junipero they are elderly and ailing. In hospital, Yorkie lies in bed on life support. The room is lit with a soft, pink-white glow, the hissing and beeping of a ventilator a companion to the slow beat of her heart. Time in San Junipero is rationed; those who spend too long in the virtual world, it is said, are driven mad by too much estrangement between body and mind. But Yorkie lives as part-machine already; she is quadriplegic, hospital-bound, and has spent decades in a medically induced coma. Is it sleep, when Kelly plugs in and visits her girlfriend? Is it waking, or living? Is it surrender to a limited, 'known' synthetic world, therapeutic and commercial, instead of the unknown sleep that is death? In real life, Kelly becomes increasingly ill. We see her wearing a nasal cannula, pensive and staring out to sea. Yorkie begs Kelly to stay, to 'pass over' into San Junipero for ever. She isn't sold yet; she doesn't want to 'spend for ever somewhere nothing matters'.

Years later, San Junipero's final shot remains as vivid in my mind as when I first saw it. It's of a data centre filled with thousands, maybe millions of glittering lights. Each represents a person, and an afterlife. A machine passes through the halls of this digital heaven, checking up on them. They are nodes now, watched over by machines for eternity, or something like it. They've achieved the best sleep, the ultimate sleep. Through technology, they'll sleep – and live – for ever.

Beyond its heartbreak, its powerful music and the immensely charming lead actresses, what I love about 'San Junipero' is its inversion of a familiar science-fiction trope. We often see people in an artificial sleep, watched over by machines designated as their keepers. The difference is that this set-up usually goes wrong, or is weighted against the protagonist (think of *Alien*, *Aliens*, *Prometheus* and *Moon*), whereas here we're given a happy ending.

Consider *The Matrix*, where people are 'harvested' for heat energy with umbilical wires, their brains connected to simulated reality with a headjack. In *2001: A Space Odyssey* the crew of the *Discovery One* travel in induced 'stasis', a form of suspended animation akin to a coma. In both of the above, machines take advantage of humans while they sleep. While they rely on a battery, computers don't wear the symptoms of their fatigue so clearly. Their bodies are not fallible like ours, especially in space; where oxygen and food are rationed, robots have an unfair advantage.

In science fiction, characters often sleep in pods instead of beds. Their sleep is more extreme, lasting for days or even years, and slowing the body's function for the sake of space travel, life extension and conservation of resources. Entering the pod, they consent to be watched over by machines; the pod is like the womb, while the robot, motherless, tends to their bodies and monitors their dreams. The pod is curative but claustrophobic; in *Prometheus*, an automated 'Med Pod' aborts Dr Elizabeth Shaw's alien embryo with barbarous precision, using lasers to cut her open before closing the incision with metal staples. In *Interstellar*, astronauts rest in 'hypersleep' in a sarcophagus-style 'hab pod', which slows down the process of ageing during interdimensional travel. The sleeper must first select a 'waking date' for reanimation, then climb into a bag, suspended inside the pod in a heated liquid which protects them against radiation and other threats. A pod, then, is a living death, a wifi-enabled coffin. It's a visual warning of the dangers of travelling too far from one's own bed, the safest place on earth.

Which is more accurate, the positive or negative portrayal of sleep with technology? Is it a matter of bargaining – you suffer if you take on the machines, but you're granted immortality if you surrender in full? While we evolve strategies to avoid it, and technologies to perfect it, no living person is stronger than sleep. For simple, evolutionary reasons, to sleep beside someone – or something – requires a degree of trust.

What does it mean to sleep with technology? Does it imply we trust our computers more than we trust other people? Community, and art, and possibilities emerge at this crossroads between technology and sleep, but we also risk compromising what makes us human. We are not batteries, and sleep is more than just a nightly chance to recharge. It's a communion, a ritual, a road to the unconscious and the phantasmagorical. It allows us access to part of ourselves that we cannot plan for, and cannot control.

In science fiction, the sleep watched over by machines is called 'stasis', 'cryosleep' or 'hypersleep'. The implication is that these prolonged, enhanced sleeps are a counterpart to prolonged periods of energy; once woken, or reanimated, a hero is ready to fight for as long as is needed, because heroes never get tired. These prolonged sleeps fly in the face of nature, and they rarely end well. Remember one of the final scenes in *2001*, with Bowman in bed, surrounded by sinister, anachronistic furniture. He's in bed, *a* bed, but he's not safe, nor is he on Earth. What's going on? Bed is a return. Bed is giving up – a dream, a hallucination. Where we're going, we won't need beds.

Drug: Seroquel
Exp. year: 2019
Dose: 25mg
The doctor preferred that I take this rather than Valium – clearly, they've taken all the fun out of being sane. Seroquel is technically an antipsychotic. When I first realised this I was a little disturbed, almost as disturbed as when I read

that one of the side effects is weight gain. But it also induces sleep, and one bad day, I finally decided to try it. A tiny, quiet sinkhole opened inside me, into which my problems disappeared.

On Christmas Day, my family and I visit a hospital in the suburbs of Dublin. A relative is staying there; he's been there for almost a year, suffering from several different illnesses, some of which seem to have been caused by the hospital itself. The entrance hall is busy; families pass by carrying bags of presents, food and flowers, and people sit outside the Starbucks stand drinking coffee. It's only upstairs that things are silent; we take the lift up to the intensive care unit, where the halls, the wards and the communal rooms share an eerie silence, unmoved by the day of the year or the people downstairs. It doesn't feel like we're breaking it with our intrusion, although we are the only people walking through the hall. Rather, we are being swallowed into the quiet, into the cold white walls and the absence of life.

I notice each room as we pass. Men lie in beds, most of them sleeping. In one or two of the rooms, though I try not to stare, I see patients with eyes half-open. There aren't any radios playing, nor are there TVs, apart from one room where I see two children and an adult sitting on the edge of a bed, watching a gigantic screen in front of them. The sound is so low as to be inaudible. The screen has a blue tinge, which seeps out and invades the room with a dim, melancholy colour. The family is facing away from me; I can't tell if they're sad, or happy, or if they're so absorbed in what they're watching that they have forgotten to switch on the lights.

Most of the hospital rooms are empty, apart from the patients and the machines watching over them. Gradually I notice that the ward isn't entirely silent; there are no human voices, but I can hear the machines in an uneven chorus. Once I notice this ambient drone, it's impossible to ignore. Beds are surrounded by ventilators and electrocardiogram screens, hydraulic lifts, suction pumps and monitors. Their hisses and beeps form a soundtrack; while the patients sleep, the machines stave off the void created by their silence.

Hospitals have historically been plagued by noise, which slows the healing process and disturbs patients' sleep. Florence Nightingale warned against it in her 1859 book, *Notes on Nursing*, and research shows that the level of noise in hospitals is rising; one 2005 study found the average noise level can rise to over 100 decibels at night, the same level as a chainsaw.[5] Such is the machine noise in the average hospital that nurses and staff report 'alarm fatigue', unable to distinguish between background noise and sounds which are necessarily urgent.[6] The sound becomes a symbol of a world spinning out of control, a world crowded with pain suffered in human silence, in halls populated with unfeeling, unerring mechanical carers. Patients remain in bed, unmoving, somewhere between waking and medicated sleep. Like the sleep pods, and hypersleep in science fiction, they are flung upon the mercy of machines.

My relative didn't even want to listen to the radio, when we saw him. He couldn't read the newspaper, or magazines, or books, because he was so heavily medicated and tired. We stayed a while, then departed, leaving him alone.

I've rarely felt so disturbed as I was by the sight of my relative in the hospital, distant and quiet. Illness is quotidian, inevitable, and I am not afraid of the body's frailty. But I am afraid of loneliness, of an interminable life spent immobilised in the presence of machines. I am afraid of the unliving mocking my lifelessness.

Where is the pursuit of the best sleep leading us? Is this science fiction made real, in a hospital ward – a connected sleep, monitored and induced?

Technology has a symbiotic relationship with loneliness; the more lonely we feel, the more we look to technology, and the lonelier it leaves us in turn. The best sleep is only temporary, and comes at a cost. The search for the best sleep will last a lifetime, and when we wake we will still be alone.

Drug: Anxiety
Exp. year: Ongoing
Dose: Unlimited
The strangest sleep of all comes when I have a deadline, or relationship problems, or when I'm angry, or stressed, or waiting for news. Dreams, talking, shadows at the end of the bed; I reach a point where sleep is no longer a break from myself, and the night stretches like a canvas, blank, waiting to be painted with fears.

What a time to be asleep. The more I read about it, the more I came to think of sleep as a fad, a cultural practice, a dreary obligation.

Once you abandon the idea of a 'natural' sleep, sleep becomes at once luxurious and optional. Technology brings us to a new frontier of sleep; it bears testament to sleep's outer limits, and our attempts to control them. We have always welcomed capitalism into our dreams. We have always slept in public. It's not enough to mediate our waking hours; now, as we sleep, we are watched over by machines.

Increasingly, sleep has less to do with the body and more to do with how we interact with capitalism. Technology aids this change, quantifying sleep, connecting us, reminding us that we're not alone even as we drift into solitary oblivion. We are lonelier than ever, and that need has created a technology of loneliness.

We are anxious, and distrustful of the body, which is always in juxtaposition with technology.

For a while I received emails from the nap hotel – I forgot to take my name off their mailing list. One especially memorable message advertised a new attraction: professional cuddle therapists. Nordic Cuddle is a 'concept' costing £65 per hour, designed to relax and aid sleep. Their website reads:

> Our sessions provide a safe space for clients to experience comforting cuddles, and to speak openly about issues that are on their mind. We listen carefully and seek to reassure, while also using the power of platonic touch to boost your mental, spiritual and physical well-being.

Even for writing purposes, I'm not sure I'd be prepared to spend £65 on being cuddled. I'd sooner invest in a weighted blanket. The idea strikes me as dystopian, but honestly so; they cut to the chase by offering a simple and necessary thing that technology can't give you. Perhaps it even works. A testimonial on their website reads 'I can't remember when was the last time I have been as relaxed as after a cuddling session. It meant so much for me to be able to have someone actually willing to hold me. Thank you.'

Studies show that women in long-term, stable relationships find it easier to fall asleep, and wake up less during the night, compared to single women. They experience feelings of safety. They have lower levels of the stress hormone cortisol, and higher levels of the love hormone oxytocin. They even tend to live longer, and have better health. Responsiveness, too, is critical; 'restorative sleep' is helped by being close to someone who understands you and who knows your needs.[7]

I think I would give up a lifetime of Netflix to be held, even for a short while, before I fall asleep. It would be a fair trade. It feels infinite. Only now it occurs to me that I have been living on outer limits, that I've been missing something I didn't know I needed. It seems to me that I have been searching for the experience of human interaction in fragments, in the way that online services 'know me', in the small, satisfying burst of attention provided by emails and tweets, the reassuring voice in a hypnosis video.

We live in a time when work, power and one's validity as a thing that walks the earth is forever being threatened by machines, and sleep offers little escape. Instead we invite computers into the dreamscape, to whisper to us in darkness. Surveillance begins to feel like luxurious flattery; we train these devices to know us, and it hardly feels like work. We sleep as guinea pigs in a global sleep lab, one where the results are sent to someone else as data, making them richer.

The internet doesn't know us, not like other humans do. It's time for me to look elsewhere, to leave the screen, and to step into real life.

PART 3

Cyborg Heart

Men Explain the Apocalypse To Me 1: First Dates

IT'S AUTUMN 2018, and I'm sitting alone in my childhood bedroom, waiting for signs of the eschaton.

Apocalypse predictions come and go. This year has had plenty: Nostradamus promised World War III, global economic collapse, a series of earthquakes in America and the eruption of Mount Vesuvius. Another apocalypse prophet, David Meade, the author of books including *Rapture 2015 and Planet X*, *Planet X: The 2017 Arrival*, *The Coming Clinton Economic Collapse* and *The End of Days: Planet X and Beyond*, predicted that the rapture would take place in 2018, shortly after North Korea became a superpower.[1] Meanwhile two asteroids – one the size of a jumbo jet, the other the size of a car – narrowly passed Earth in September, at a distance closer than the Moon, but did not collide.[2]

Now it's November; six weeks remain for the predictions to come true. If the apocalypse really is about to take place, then I don't want to be raptured alone.

Over the last few years I've had several short-term relationships, rarely lasting longer than a few months. Those fell away, pleasant enough experiences that nonetheless left me feeling more alone than before. The problem has been that in the three years since he broke up with me, I have thought about my ex every day. It feels like I have no future without him; I think of him constantly, and these thoughts have coalesced into a layer between me and the world, preventing me from forming a meaningful connection with anyone else.

It hasn't helped that he keeps writing me emails. I speak to him almost every day, but he doesn't want to date me; he wants to exchange hundreds, perhaps thousands of messages, and to keep me at arm's length. This, too, is a pleasant enough relationship; my ex no longer lives in Dublin, but I know he's there, a name beside an 'unread' notification, ready to talk about films and books, politics, cultural theory, and the apocalypse. We talk a lot about that.

Back in 2016 we were only together for a few months, but now years have passed. In one way he restored my faith in technology as a way to bring people together. In another, I interpret his messages as a sign that I'll die alone, after dedicating years to a spectre in my inbox. The love I have for him – and I know it's love, it has been long enough – is unwieldy; I have so much, I don't know what to do with it. It would be a terrible thing to let it die inside me, without reaching its target.

I feel like a nun wedded to the internet. I pour love into writing the perfect email, night after night, uncertain if he will ever love me back. Sometimes I wonder if it's this system I love – the writing, the mediation, the ritual and the safety of distance – rather than the person who receives the message. Our relationship – perhaps not a relationship in a conventional sense, but a kind of connection, still – developed in the years after I was given my mental illness diagnosis, a time when, on some level at least, I concluded that I was unlovable, and unsuited to serious relationships. I resent the ambivalence between us, which email leaves permanently open, but at times I also find it comforting. It provides the possibility of love, without the need to take responsibility for it.

In my parents' house, in Dublin, it feels like a lifetime has gone by and I have nothing to show for it. I know that I'm frozen in obsessive patterns; I need to tell him how I feel, or move on. I need to connect with someone in real life, and step away from the screen.

Winter closes in, a landscape of hot whiskeys and parties, anticipation, loneliness and cold. My mind changes; I begin to believe that I can compartmentalise my dating life, channelling love into the daily emails while finding more casual relationships through apps. I decide that going out with men from the internet will be good for me. It might even take my mind off my ex.

I decide to go with Bumble, billed as a dating app for feminists. It was formed in the aftermath of an employee schism, the kind that's inevitable two or three years after a wildly successful start-up is founded. Facebook had Eduardo Saverin, who sued and settled with Mark Zuckerberg for an undisclosed sum in 2009, and Tinder had Whitney Wolfe Herd, who sued Tinder for sexual harassment in 2014. In 2014, Wolfe launched Bumble as a competitor to Tinder, promoting it as a dating app where women made the first move.

In the past I did well with OkCupid, another, older dating app. Years before, when I moved to London after college, it was how I ended up making friends. I tried Tinder, but didn't last long – I found I could only stand to use it when drunk, or on Valium. More recently I lasted for an hour or so on Feeld, an app aimed at 'alternative' lifestyles like polyamory, which I had also tried years before with varying results. In my early twenties the prospect of a jealousy-free relationship appealed to me; I ended up dating a guy with several other girlfriends, but found it jarring when I eventually, inevitably met them, and perceived myself suddenly as part of a menu. We were all quite different – one was in the sciences, another literary and bohemian, and I was the weird internet girl from Ireland. Yet we were all variations on the same thing: educated, middle-class, and drunk on our own daring at engaging in this unconventional form of affection.

None of these options seem right for Dublin, because the city is small. This becomes clear very quickly on my first night using Bumble, when I run out of swipes within the first hour, and men who have paid to upgrade their accounts start to reappear on my screen. I know I've gone back to the start when Odysseus appears again. His profile says he's thirty-nine, but somehow he looks older, and his reappearances – three times, I think, in one night – make me nervous. I find it slightly sad that a man named after an ancient hero is here, adrift on a downloadable dating app.

Still, I find Bumble easy to use, and increasingly addictive. It's less sleazy than Tinder – a cocktail bar, rather than a nightclub for college freshers – and I quickly notice that most of the men on the app have jobs with well-known multinational tech companies. I have deliberately set my age limits higher than my own age; I'm hedging my bets that this way there'll be a lower chance of fuckboys. There's a large number of men with foreign-sounding names, with flags in their profiles indicating where they come from – Canada, the US, Brazil, Japan, France and Spain. I wonder if Bumble makes Dublin feel like a more cosmopolitan place, somewhere with possibilities, the way OkCupid did in London. What OkCupid did, back then, was reveal that the people around me were human; they had quirks and opinions and funny pictures of themselves, favourite films, witty retorts to questions the app routinely asked its users in a bid to get them talking. Online dating made me less afraid of the anonymous professional hordes in their black coats on the Underground, the gleaming, intimidating strangers I passed every morning on my way to work. Maybe it can also make me see Dublin in a new light.

I get lucky. I get a lot of matches. The two men I decide to send messages to are generically good-looking, preppy, with dark hair and a beard. Both work in tech, and both are in their mid-thirties. They are so similar I almost instantly forget which is which. When one man responds I'm not sure which one he is, and I realise, with an exhilarating chill, that I don't particularly care either.

I read through the bio for the guy I'm talking to. It says he's French, and that he works at one of the biggest tech companies in the world. He's in Dublin for meetings, most of which involve alcohol. We end up talking over messaging about linguistics, and tardigrades, the microscopic creatures known as 'water bears'. He asks me out for drinks. I arrange to meet Apocalypse Date #1 the following night.

We start at the Hacienda, a beautiful, deeply weird pub just off Capel Street where you have to ring the doorbell to get inside. It looks like the past, like the Black Lodge from *Twin Peaks* if it was decorated by an Irish granny. I figure the place itself will be something to talk about; it's meticulously cluttered, almost every inch of wall covered with small, framed pictures and paraphernalia stacked on every surface. On the bar there's an antique diving mask and a collection of oil lamps. We both order gin and tonics, then we find a table.

He tells me about his work. He tells me about European cities, about the Fermi paradox, about a means of production that's broken beyond repair. I stay as quiet as I can. Part of the appeal of this experiment is the chance it gives me to get to know someone without revealing too much about myself. Ahead of the date I have promised myself that I won't talk about writing, about living with my parents, or about being certified as emotionally unstable.

He tells me about the apocalypse. His interest is in 'collapsology', the name given to the study of the slow end of industrial civilisation. I am still coming out of a phase in which I nurtured an intense crush on the Unabomber, so I listen, even though it's hard to take seriously coming from someone who works for the technocapitalist menace. Reading about collapsology later, I find a video in French; an interview with a handsome, similarly bearded French man identified as a 'Collapsologue'. It's Pablo Servigne, an author who along with Raphaël Stevens wrote *Comment tout peut s'effondrer* ('How Everything Might Collapse', 2015). Their book envisions the end of the world in our lifetimes, based on predictions from disciplines including anthropology, biogeography, biophysics, law, ecology, economics and art.

Men have told me in the past that I'm intense, but it occurs to me now that we've jumped straight to a conversation so earnest, so serious and *so* intense that it borders on parody. Perhaps we're doing this to prove we're serious people, that we're capable of looking beyond flirtation, the visual cues we're absorbing even as we talk, and the apps we browse through looking to find someone to sleep with. Perhaps the apocalypse, here, is a signifier, a shortcut to a 'deep meaningful conversation', so that we can tick a box and prove we're both capable of that, too, before we have sex.

Collapsology has been criticised for what the eco-socialist Daniel Tanuro calls its 'fatalistic resignation', arguing that it focuses too much on 'grieving' and not enough on addressing environmental and economic problems. Another critic, Nicolas Casaux, called it 'nebulous' and said its main problem is 'the narcissism it perpetuates', accusing its supporters of dwelling on apocalypticism and bearing 'toxic hints of the dominant culture, which prevent them from taking a more determined position'.[3] Apocalypse Date #1 seems to move between this same resignation and a more practical approach. He tells me he has become a vegetarian for environmental reasons. He would be vegan, except that he really loves cheese. I sympathise. I have been a vegetarian since I was eleven years old, but in recent years I've found it hard to keep track of the reasons why. I don't know if I really care about animals, or the planet, as much as I suspect I should. Perhaps I pursued vegetarianism simply because I had multiple eating disorders, and it helped me to restrict. As it stands, today I'll eat bites of other people's burgers, or pieces of sushi, or free samples of Marks & Spencer's steak pie, but I never buy meat. I eat eggs every morning, too, but beyond that I am basically vegan, sort of. But all of this is difficult to explain, and risks making me sound completely crazy, an area one flirts with, I find, in speaking favourably about veganism, unless you're a handsome and accomplished bearded man who possesses gravity and an attractive French accent. Then people, especially women like me, will take you seriously and nod in agreement as you speak.

I nod in agreement, but I can barely hear what he's saying across the table, through his accent and the background noise. After five or ten minutes I know that I'll definitely have sex with him. It's enough that he spends his time thinking about annihilation, that apocalyptic fantasies live in the mind of this very clean-cut, gainfully employed person – the opposite, I feel, of myself, with my precarious work, my second-hand leather jacket and stretchy black tube top that I'm wearing as a skirt tonight. Clothes are the main way I demonstrate my feelings about the end of the world, a world I see filling up with mass-produced non-recyclable trash, and cheap clothes that fall apart after one wash. All they have is the first night you wear them, then their life is over. I demonstrate my anxiety about the environment by buying clothes almost exclusively from charity shops, vintage shops, flea markets and eBay. I like things that have had at least one owner, and one life, before they become part of mine.

It turns out that Apocalypse Date #1 is a fan of *Rick and Morty*, the darkly comic, sci-fi Netflix cartoon. He and I have that in common. His favourite episode is 'Rick Potion No. 9' – season one, episode six – in which Rick accidentally 'Cronenbergs' the world, transforming humans into grotesque mantis-hybrid creatures before abandoning his dimension for a new one. Together, Rick and Morty find a new dimension where their alternative selves are about to die. Once this happens, they bury their own bodies in the family's backyard. Date #1 tells me that this plotline helped him to make peace with his own mortality.

I tell him my favourite episode is also the one that fucked me up: season two, episode three, 'Auto Erotic Assimilation', in which we finally meet a love interest for Rick. Unity is a curvaceous hivemind intent on assimilating the universe in its entirety – Rick is, quite literally, in love with an apocalypse. The only person Unity values as an individual, it appears, is Rick, with whom she embarks on a planetary orgy involving hundreds of assimilated and willing young redheads. Recognising that this can't continue, and that Rick, and her love for him, will get in the way of her mission, Unity ends their relationship. Afterwards Rick attempts suicide, using a machine of his own design, but he's so drunk that he passes out and the laser misses his head. It's a desperately sad scene, soundtracked by a song I've played so many times that I know the lyrics by heart. Someone has even made a ten-hour-long remix on YouTube. The song is titled 'Do You Feel It?', by a band called Chaos Chaos. The lyrics are about searching for immediacy, honesty and love in a world that seems to be slipping away.

After the Hacienda we go to the Liquor Rooms, where I show him the taxidermy bear at the back of the bar. I always get self-conscious while ordering here, because the bartenders are employed, it seems, to perform a spectacular version of bartending – one rooted in a fictionalised version of the past, as much as their job is to sell us drinks. They wear their shirt-sleeves rolled up, with sleeve garters and braces, and they perform an elaborate dance of shaking the mixer and straining it each time. I feel like it puts pressure on me; I don't want to pick a drink that causes them extra trouble. Nor do I want a drink that has egg whites in it; that seems, to me, just wrong.

The menu has been reimagined since my last visit as a tribute to famous Irish women. I find it a little try-hard. I choose 'The Gazelle', a drink for Eva Gore-Booth made from cognac pear liqueur, soda water, pineapple, rum and something called 'ginger shrub', although, when it arrives, there is no visible shrub in the glass. Apocalypse Date #1 discusses neurochemistry, 'flow state' and the science of happiness. He dances around the topic of his work, and the internet in general, explaining how social media apps hijack the chemicals in our brains – dopamine, in this case – to create addictive loops of anticipation and short-term satisfaction.

The drinks are very strong. In the middle of a second one, the song from the *Rick and Morty* episode we talked about earlier plays over the sound system. I'm surprised; I've never heard it played anywhere apart from my laptop. I tell this to Apocalypse Date #1, and he asks if I'd like to go back to his hotel.

On the walk along the Liffey I watch as he encounters parts of Dublin for the first time, and wait for his reactions. He tells me he was expecting it to be different, that the city is patchy and dirty and run-down. He's right; there are parts that are ragged, buildings left empty for years next to cranes and the glassy headquarters of tech firms. I tell him about the average rent in Dublin, €1,620 a month, a record high. For years, Dublin, where I grew up, has experienced a housing crisis. It's been going on for so long that it has stopped seeming like a 'crisis', implying something sudden, and has begun to feel more like culture instead. The government has failed to deliver new houses, tenant rights have been eroded, and rents have steadily risen to 37 per cent higher than they were ten years ago.[4]

I'm quite drunk now, but I attempt to explain to him that I spend hours sometimes on Expatistan.com, a site that compares international living costs, where Dublin ranks as the twenty-fifth most-expensive city in the world. I even browse rental ads in other cities, contemplating a buffet of alternative futures.

He's staying at the Marker, an expensive hotel in the Docklands that I have mocked several times before in writing. The outside is paved with mirrored glass made to look like waves on the canal, and the building is set in a concourse filled with odd geometric shapes, red sticks of neon which emerge from the ground, and the scalene twist of the Bord Gáis Energy Theatre. The area is calm, and empty. I'm happy that there's no one around to see me going inside.

The lobby is silent and unoccupied. I wonder if we'll kiss in the elevator, but we don't. We pass through dim, wood-panelled halls that seem to retreat into darkness, into tastefully lit oblivion on both sides. His room has a purple carpet, a grey bedspread, and a picture of the two chimneys at Poolbeg on one of the walls.

I like the stillness of hotel rooms, the neutrality. The only times I stay in hotels is on work trips of my own. We lie out on the bed side by side. He signs in to his laptop, and I read his full name on the welcome screen, trying to remember it to stalk him online later, but I almost instantly forget. We begin to watch one episode of *Rick and Morty*, then we have sex. The next morning we walk out of the hotel, say goodbye, and he walks straight into the offices of a nearby tech multinational.

In the days afterwards I feel like I am under a spell. I consider messaging Apocalypse Date #1, but don't because he hasn't messaged me first, even though this goes against the 'female empowerment' message of Bumble. To try to contact him now seems inappropriate, as though I am 'catching feels', and view him as more than a one-night stand. Feelings, like love or even mild genuine affection, seem hostile to dating apps. I reserve my feelings for emails instead.

Date #1 didn't try to conceal the fact that he was only in Dublin for a few days, and that our time together, ultimately, would be meaningless. This is why I begin to think of him as an Apocalypse Date – not only for the subject of our conversation, or because of the prophecies made for this year, but because tomorrow the slate will be blank again. He'll be in another city, and will take out his phone and swipe for someone new. He inhabits a world where old behaviours have been disrupted and rebuilt by technology. It's a little dehumanising, and cruel, but also alive with possibility.

I remind myself that Bumble is full of interesting people and experiences, and that I too should continue swiping for more. Affirming this thought, Bumble persists in sending me messages; it says, 'Hurry up! Say something, a connection is about to expire.'

Tardigrades, the millimetre-sized creatures I discussed with Apocalypse Date #1, are often referred to as the animals most likely to survive an apocalypse. They're rotund and knobbly things, halfway between microscopic plush toy and origami structure. They're a pinkish-brown colour, fleshy, with tiny claws and a mouth resembling a rubberised floral anus. They're sometimes described as 'water bears', and are considered almost indestructible.

Tardigrades have survived in extreme climates, at the deepest parts of the ocean, and for long periods without food, water or oxygen. When dehydrated, they can survive for up to 100 years, and in 2007 they became the first animal to survive exposure in space. They're known for their ability to colonise new and extreme territories, paving the way for further life. What people talk less about are the tardigrade's mating habits. In 2016 a group of researchers at Germany's Museum of Natural History produced a tardigrade sex tape, filmed under a microscope. The male tardigrade wraps itself around the female, and is stimulated by the female tardigrade's 'mouth-like opening'. For about an hour there's intense, messy tardigrade foreplay, and the male tardigrade ejaculates several times.[5]

My opinion is that the tardigrade should be an icon in our apocalypse-fixated culture. After the world ends, it's likely there'll be cockroaches, zombies and tardigrades, and personally I'm rooting for the tardigrades. There has been debate over whether tardigrades are truly extremophiles, creatures capable of thriving in physically or geochemically extreme conditions, or whether they simply shut themselves down to resist their surroundings. In one interview, an evolutionary biologist named Jim Garey, from the University of South Florida, said 'Tardigrades are not true extremophiles because they are not adapted to live in extreme conditions. They can merely survive exposure to such conditions. The longer they undergo such exposure, the greater their chance of dying. Tardigrades are always waiting for something better.'[6]

I, too, am waiting for something better. There's a point in the plots of books and films and TV shows that I've always found enjoyable, right before two characters you know are going to end up together kiss for the first time. Usually I don't have much interest in romance plots, but in these moments I find myself completely absorbed. I love these moments in real life, too, when I hook up with someone after a long wait. It's something you can't replicate; you can do it only once. You could almost say it's better than sex, or, at least different, because the sex hasn't happened yet; it's implied, but it's still in the imagination.

What Bumble offers its users is that same wait, again and again, in mechanised anticipation. They know what they're doing sending all those push notifications. They're selling possibility. I fall for it; I am lured into extreme conditions, in the hope of adaptation and survival. I go back and keep swiping, in search of Apocalypse Date #2.

*

Ten days later, at the end of November, I match with another guy on Bumble. He and Apocalypse Date #1 have a few things in common – he, too, has dark hair, and a job at another, different tech multinational, a rival to the one Apocalypse Date #1 worked for. In the interim I have swiped and matched with a number of guys, and almost all of them work for tech companies. I can remember only one who did not; he worked for a charity, and in his profile he listed his ideal dinner guest as 'The Donald'. I matched with him out of morbid curiosity, expecting a 4chan troll. He turned out to be surprisingly sincere – in his manner, as well as in his support for Donald Trump – and I moved swiftly on.

Apocalypse Date #2 comes from South America. He strikes me as earnest, ambitious and a little sad. After one drink he tells me he recently broke up with a girlfriend, and I realise that the aura of her hangs over him like a transmittable illness. It's jarring, because until now I've thought of Bumble as an orgy, an exuberant place filled with resilient singles (and, I imagine, cheating husbands – in fact, it occurs to me that Date #1 was exactly this, but I try to dismiss this thought as paranoia). We meet at the Bernard Shaw, a bar I've been going to since my fake ID days, where they've opened a food market called the Eatyard. They're selling American state fair-type foods, animal-style chips and cheese fondue. The bars, covered in fairy lights, sell variations on hot whiskey. It resembles an alcoholic Santa's Grotto. We buy drinks – I get a gin and tonic, again, and he has a beer – and we sit outside under a wall of graffiti, where his friends are passing around a joint.

I've been on these dates before, the kind that at the start aren't really dates at all, but an invitation to join a guy with his friends on a Friday night out. If we get along, then at some point we'll break from the group and go somewhere more private. At face value these encounters are easier than a one-on-one date, but in reality they're harder. They're a very casual vetting process, conducted by a group rather than an individual.

It's hard to talk to Apocalypse Date #2. The bar is loud, everyone around me is talking in Spanish, and it feels very awkward to interrupt them. When we talk we aren't hearing each other, only talking to ourselves and nodding vaguely when the other person opens their mouth. After a while it occurs to me that he knows I cannot hear him. I sit and look around me, watching groups of people so young they must be college freshers, or, like me when I first started coming here, still in school. There are older people here, too, hipsters of the mid-2000s vintage, the kind who once wore American Apparel deep-V T-shirts and disco pants. Certain Dublin venues – the Bernard Shaw and Whelan's come to mind – have maintained a stasis since that era that can only be deliberate, playing the same music and serving the same two-for-€12 mojitos. Perhaps this was already old when I first went to these places as a teenager. Perhaps culture in Dublin has stayed the same for multiple generations.

Three joints in, I realise I'm not feeling much apart from a vague gluey-ness in the space between my brain and my mouth. I've become less articulate and more reticent, a quality I often wish I possessed while sober. Perhaps this will be an advantage to dating men whose first language is not English; I won't talk too much, and even if I do they won't understand me. For years now I have known that I talk too much; I have that egomaniacal quality. Often I do it for no reason other than out of a fear of silence. I fetishise mob wives, people who can take their secrets to the grave. Dublin is too small a place to talk as much as I do.

Eventually his friends move on to another place, and as with every time I've met someone's friends while on a date from the internet, there's a vaguely salacious undercurrent as they say goodbye, like I'm being read. Like they're speculating on whether we're going to have sex later, and whether I deserve it. Or whether *he* deserves it; I'm never quite sure. I have no idea what the girlfriend he recently parted ways with was like, or whether he regularly meets internet dates and introduces them to his friends. As with the app itself, I am one in a deck of cards, waiting to be taken off the stack and placed – swiped – to the right, or the left, even after we've matched on screen.

We talk in a cursory way about ourselves. I tell him I write, but not what I write about. I tell him I've lived in Dublin and London, and a few other places for short periods. I resist telling him that I live with my parents. He tells me he's been in Dublin for four years, and has worked at two other tech multinationals, both of them household names. I didn't realise people did that, drifting between them. I thought you had to swear allegiance. He tells me he's about to leave the one he's currently at for yet another. Then he tells me about where he comes from. He talks about crime and food shortages, and how he studied medicine at university and the lab was constantly broken into. He tells me about being mugged at gunpoint, about carjackings and political unrest.

I nod and don't say anything. I feel bad for not keeping up with international politics, but more than that, I'm worried that anything I say will be a generalisation, spoken in ignorance. Friends of mine talk a lot about South America on Twitter, dismissing press coverage as untrustworthy, or a psyop, or part of the US imperialist agenda. There are days when my mental health isn't all there, where I read their tweets and descend into paranoia. It feels like no source can be trusted, like everything is mediated and unreal, and slowly collapsing in on me. When I feel that coming on I try to stop it; I sign out and go for a walk, and try to establish contact again with the world around me.

But that's a lot to explain on an internet date. He has a lot, I sense, that he's leaving unexplained too.

We go for chips in a pastiche American diner, a restaurant with a spare yellow-and-red interior lit by neon signs on the walls. It's familiar in a cartoonish way, like an Edward Hopper painting finished by animators from *The Simpsons*. We share fries with chilli sauce. I decide it's not technically breaking my vegetarianism if I avoid the chunks of meat in the sauce and just go for the chips, while he talks about his visit to the Web Summit in Lisbon. I already feel tired. My voice is hoarse from smoking, and from trying to be heard over the background sounds of the Bernard Shaw. He tells me about how during his medicine degree, he went into the lab early one day and found a freezer box. He looked inside, and discovered that it was full of human heads. He tells me that afterwards he stopped studying medicine and switched to business studies. 'I am a capitalist,' he says, the first time I've ever heard anyone say that to me in conversation.

There is not an ounce of sexual tension between us, but he tells me he lives a few minutes' walk away, and I agree to go back with him, unsure of what will happen. It is only when we get to his flat that he reveals he loves *Rick and Morty* too.

We sit at his small kitchen table and watch Netflix, the episode titled 'Get Schwifty', where a giant head threatens to destroy the world unless the inhabitants of Earth perform pop songs. His flat is a one-room studio with an en suite bathroom carved out of one corner. It's tiny, but very well located. I know, from the time I spend on Rent.ie and Daft.ie, that a room like this one easily costs over €1,000 per month. I've been curious for years as to where the start-up multinational employees live; for some reason – probably poverty of imagination, or lack of empathy – I imagined they all had homes in the Docklands, big spacious flats in those clinical apartment blocks the developers struggle to fill, among the NAMA towers. Where Apocalypse Date #2 lives is unexpected. It's cramped, and unimpressive, but still unattainable to someone with a job like mine.

We smoke a fourth joint, and I know that by tomorrow I'll be crackly and hoarse, and potentially in pain, but I smoke it with him anyway out of curiosity to see if it will make me attracted to him. His bed is two feet away, but I'm almost completely certain we won't have sex. I feel distant from all this, my movements drowsy and weighted. Ours feels more like a fraternal relationship, like I'm someone he can talk to when he's tired. He talks about anxiety, and about stress at work. I've read about the workplace environments of multinational tech companies; they have high staff turnovers, and high rates of exhaustion. Most people stay late at night, some even work at weekends. I wonder why he keeps seeking out the most stressful places to work. I wonder also why he keeps smoking weed. I used to smoke every day, believing it to be therapeutic, but all it did was eventually bring on a nervous breakdown.

It's 3 a.m. and I wonder if I'm dissociating from my own body. I'm barely contributing to the conversation, barely capable of the facial expressions required to signal that I'm entertained. I fall slowly into my own chair, and force a laugh at a second episode of *Rick and Morty*. My voice sounds far away, tinny and unfamiliar, as though coming through a radio. It seems like a waste of a date not to have sex, though I know I didn't want to. I wonder if, as a woman, I should think this way.

I tell Apocalypse Date #2 that I'm going to go home. He walks me to the door and we hug goodbye, and I know I'll not see him again. Weeks later I check on him again, on Bumble. It reads 'DELETED USER', and I wonder if that means he's back with his ex-girlfriend.

I spend December not sleeping in London. Occasionally I open Bumble on my phone, but find it hard to summon the enthusiasm necessary for starting conversations with strangers. Dating apps in general start to lose their appeal; instead I exchange emails with my ex, even more frequently than before, and start to entertain the idea that he might feel the same as me. I still can't bring myself to ask him directly; we exchange news, ideas, opinions, everything apart from how I really feel.

I fly back to Dublin before Christmas, and go to a party on New Year's Eve. It's small and friendly and nearby, and not at all crazy, a gathering of men in their mid-thirties. I am younger, and the only woman there.

It's the first New Year's Eve in a few years where I've not been in a relationship. It's also the first in a long time where I won't stay out all night, or drink a lot, or take MDMA, or all three. I'm certain of this before the night has even started, and I find it a relief. I know I'll be able to walk home and wake up the next day and I'll be able to write. I know that tonight I'll have a bed to myself. I'll put on pyjamas and drink coffee, which I always do late at night, and browse the internet and rub oil sold by The Ordinary into my face while slowly staving off a panic attack. Then I'll wake up the next day, moisturised and alone.

At the party I sit on a sofa at the back of the living room and open a canned gin and tonic. I talk to the guy sitting beside me, who promptly asks me about what I think of the author of a well-known book about the alt-right and 4chan. I predict where this is going; he remembers that I interviewed her, and he's giving me shit, shit-by-association, because he's one of those people who likes to tweet that this particular author has grown too close to her subject and has betrayed the left. The prospect of talking about online politics gives me anxiety. I try to exhibit mob wife levels of reticence.

Midnight comes and goes. We talk, my bag of canned gin dwindles, and I reassure myself that tonight doesn't need to be a big night out, that I can go home when I want and that I don't owe anyone anything. I leave at 1 a.m. and walk home playing a podcast about serial killers out loud without my headphones, which didn't fit in the tiny lizard bag I brought along for the night.

The memory of the conversation with the guy on the sofa nags at me. It reminds me that the internet is for ever, even if the people who populate the internet also seem to long for the end of the world. It also connects with a broader set of anxieties I experience so often that they keep me awake at night. I worry that people are judging me on the internet. I worry that I've ruined my chances as a writer, and as a human, by avoiding social media and its politics in recent years, ever since psychiatrists told me to do so, and that now everyone on the internet hates me. I worry that politically I have nothing to offer, only futility and misanthropy and despair, and that people who were once my friends can see this, a dark flint of cynicism at my heart, and regard me as toxic.

Over the years I've tried to educate myself about politics in Ireland and abroad, but the more I do the more useless I feel, and the less I'm inclined to think things will ever be different. When I fall into these thoughts I try to stop them, because they devolve into self-destructive cycles. I know they're a symptom of depression, and of problems in the world in a broader sense.

As I walk home I hear parties behind doors and through windows, the voices and laughter of people I don't know. I tell myself it's a new year, with a new set of apocalypse predictions. Perhaps 2019 will be better than what came before.

The next morning I get a Facebook message from the guy I sat beside on the sofa at the party, asking me out on a date. He will be Apocalypse Date #3. Unless he has that self-destructive impulse I myself have indulged in the past, the need to fuck the enemy, it seems safe to assume my suspicions were wrong; he doesn't think I'm alt-right, or evil.

In the first week of January, I agree to meet Apocalypse Date #3 for dinner. Before that we talk online, about the Taoiseach,

Leo Varadkar, about Graham Linehan's quest to make the internet hate him, about the implosive presidential candidate Gemma O'Doherty and mostly, it seems, about professional wrestling. He has a habit of bringing up subjects that are lurid and entertaining, talking about them in detail, then dismissing them as morally wrong. I've noticed a lot of guys do this when they're talking to me, showing deep knowledge of a subject while declaring it 'problematic' at the same time. I wonder sometimes if it's because virtue is gendered in the internet era, and if men can admit to themselves, and to each other, that they actually openly enjoy talking about these things only when women aren't around.

Before the date I experience a vague sense of dread, worrying again that he's doing some kind of recon job to find out if I'm 'problematic'. We message about *GLOW*, a Netflix series which I stopped watching after two episodes because it felt cynical, like it was trying too hard to brand professional wrestling as feminist while featuring lots of shots of women in shiny high-cut bodysuits. Then he sends me a video of two female wrestlers, one half-conscious and seated in a chair. The other, a woman with long, dark, banshee-like hair, takes several steps back, then crouches and bends over. She plunges one hand into her shorts, which look like they're made from red latex, and produces a tampon, which she forces her opponent to eat. Then she leers at the crowd and walks off, as they chant, 'YOU SICK FUCK! YOU SICK FUCK!'

I'm unsure if this is a calculated move to shock me, to amuse me, or to work out if I'm into really weird shit. I reply 'Straight to Pornhub', then instantly wonder if I shouldn't have said that, and whether I'd be better off pretending not to know what Pornhub is at all. Reading over the DMs we exchanged, it seems clear to me now that I was trying too hard at putting on a 'quirky' persona. I grew up in the era of the Manic Pixie Dream Girl, and have never quite let go of imitating her, in an attempt to make my inconsistencies charming.

It also occurs to me that throughout these messages each of us was trying, subtly and cumulatively, to calculate the wokeness of the other person. What were we trying to prove? When I try to be woke, I feel myself pulled by a centrifugal cultural force; I can't separate my feelings from those I believe are expected of me by the internet.

In recent years I've tried to stay woke about wokeness itself. I want to be a good person, and wokeness isn't incompatible with that, but it sits uneasily with me as a trend, an invitation to the good-and-evil, black-and-white binary logic of mental illness. I worry about my wokeness, but I also worry about worrying about my wokeness, because what kind of solipsism is that? Wokeness feels like an online distraction, a surrogate for genuine politics. I know it's a trap, but I can't give it up entirely.

I think about this on my way to the restaurant, going through the same thought process I used to experience before meetings of a socialist reading group I once attended. We'd meet in the back room of an old-fashioned bar, like seditionaries plotting to put capitalism out of its misery, but mostly we'd end up discussing internet politics. Each month, before another meeting came around, I'd write down talking points in a notebook to bring with me. I'd look in the mirror before leaving for the meeting, and ask myself if I was dressed in a manner too feminine, too flashy, too expensive, too grubby, too ironic or too weird to seem convincingly concerned about politics. I find myself doing this again, now, before the date. Clothes are the joke I share with myself; I wear dark-red nail polish and lipstick and a ratty white fur coat, in a travesty of what I feel glamour is. I want to be feminine in the way that Courtney Love was in the 1990s, a kind of monstrous feminine.

We meet and we hug, not awkwardly but with a certain

deliberation, as though considering the possibility of being close to each other. Then we go inside. The restaurant has wooden tables and narrow benches, high off the ground, which my legs dangle from, and we consult the menu.

I find it hard to eat with other people; food is a minefield, a performance. It's residual fear after years of eating disorders, a tendency I try to pre-empt by looking up menus online in advance, and picking out the 'safe' foods. Vegetarianism complicates this further; I don't want to draw attention to myself, but sometimes the 'safe' menu option is also one with meat. This leads me to picking at food, hiding the meat under lettuce leaves or similar, and drawing more attention to myself in the process than I would if I was honest from the start.

I want to avoid this situation. I try very hard not to calculate calories. After probably far too much thought, I choose a salad, and the waiter disappears.

We talk about social media advertising, about wrestlers, and about the rental crisis in Dublin. I tell him I spent the previous month in London, and that I keep thinking about leaving for somewhere else, but can't commit to a place yet. Date #3 tells me about working for a tech multinational, another of the best known in the world. He seems critical of social media in general, but not in a decisive way. I still can't get over the fact that every man I have met, in the time since I started this experiment, has worked for or is working for a tech multinational. I'm beginning to think this is a code that I don't see, and that a significant percentage of the people around me at any given time in Dublin have worked for Google, Amazon, Microsoft, Twitter or Facebook, and I just don't notice, but that they can all somehow identify each other.

I ask questions, but he knows I'm a journalist, and I think he's a little wary of me. It's likely he was asked to sign a non-disclosure agreement by his former employers. I try not to push it, but I speculate; it's possible he only took the tech job for money, but it's equally possible that he originally believed in what the company was doing, and in the Silicon Valley ideology in general, until some personal event, or boredom, or frustration with work drove him out of the field.

I berate myself for thinking this over as much as I do, and for having stayed awake at night paranoid, convinced that he asked me out due to some strange, cynical impulse to expose me. Then I mention the TV show *It's Always Sunny in Philadelphia*, which I watch far too much of late at night, and he tells me the show is problematic because one of the writers originally dated one of the cast and fired her after they broke up. I feel at once embarrassed and resentful. By many people's estimation, *It's Always Sunny in Philadelphia* is a show with a dubious grasp on what is socially acceptable. But the cast is the least of its problems; he should probably be more concerned with Danny DeVito's character, Frank, and his treatment of women, or his collection of guns, or his past career operating a sweatshop in Vietnam.

There are plenty of good reasons to enquire about someone's politics on a first date; better to find out early on if someone is a racist, or a sexist, or a dedicated follower of QAnon, and get away from them as quickly as possible. But I don't think that's what's happening. There's a wall between us; we're talking, but more as a means of self-fashioning than as communication. Maybe there's always a wall between me and other people, especially when I'm talking to men, which leads me to question why I'm conducting this whole dating experiment in the first place.

It feels like our interactions are limited by the parameters of the screen; social media looms over our conversation, and we struggle to live up to our online selves.

After a while I realise that I like this guy but that the date is hard work, like I'm having to watch myself and that anything altruistic or good that I say will now take on the air of a performance. We go for drinks afterwards in a bar nearby and sit outside. A woman comes around and asks us for money, and I have no change. She doesn't believe me, and for a moment it feels as though it might become a confrontation. 'Sure you don't,' she repeats. 'I don't,' I say. 'Sure you don't.' Eventually she moves on. I'm certain now that whatever I do, he's going to think I'm an asshole. Over the past two hours we haven't connected; we've done little more than exhibit polished, internet-friendly versions of ourselves, self-congratulatory and cold.

I wonder if my neurosis transferred to him like bad energy, infecting our conversation, or if we are products of the same fraught climate, the same hopes and fears. I wonder why we both spent so much time and effort on convincing each other that we're good people, and if it achieved anything, and if we are even remotely attracted to each other, or to virtue in general as a sexual characteristic.

As we make our way to the Luas stop on Marlborough Street he puts an arm against my back and explains where we are, and where the station is. We are on O'Connell Street. I say that I get the Luas nearby almost every day, and he replies 'It's OK, don't worry, we're just on the Northside.' It's a joke, but it's a patronising joke, and I know now that he thinks I'm sheltered, and privileged, and ignorant, and a snob from the Southside. It's true, I'm certainly one or two of these things, but it feels like he's reaching for the last word in what has been a tacit confrontation. I don't comment, and soon after that we say goodbye.

Apocalypse Date #4 appears suddenly, late one night in January, when I'm at home working on five different deadlines. Clicking between tabs and browser windows, I decide impulsively to swipe through Bumble again.

I hardly talk to Apocalypse Date #4 before we meet; after we match on Bumble we only exchange basic details. I tell him I write, and he tells me he's studying and working in retail. Like Apocalypse Date #2, he comes from a country in South America. He likes horror movies and video games. Occasionally he uses unusual words – 'genial' is one of them – which suggest he's using Google Translate to talk to me.

When we meet for our date I am already a little drunk, arriving from a friend's thirtieth birthday party. I feel bad about this. It's definitely not respectful. We talk and it seems as though he can't hear or understand what I'm saying. I already speak quickly, but now I'm speaking extra-fast because I'm nervous, and possibly drunk. I'm wearing a black knitted dress, long-sleeved but short, with black boots and patterned lace tights. My friend described it earlier, at the party, as 'slutty 1960s', which made me laugh.

We go to the Globe, which is loud, and filled with what turns out to be a Meetup group, an online community occupying real-world space. I used to go to these groups in London; they're useful for when you've arrived in a city and want to make new friends. I drink two gin and tonics, he has a whiskey and coke, then he suggests going somewhere quieter, and we move to the cocktail bar next door.

In the ridiculous cocktail bar, where we order the special that tastes like glamorised adult Calpol, we continue to learn about each other in broken conversation. Apocalypse Date #4 is studying for a tech qualification, and learning to code in Python and C++. Of course he is. It occurs to me that I'm drinking to make myself attracted to him, to blot out the other things on my mind and to speed us towards sex. It occurs to me that this is callous, but also that it is precisely what dating apps were designed for, and that this might be what it means to be part of what an article in *Vanity Fair* once termed the 'dating apocalypse'.[7]

I consume four drinks in total, then I follow him home to a room in north central Dublin. There is no great chemistry between us, but I tell myself that there's a kind of camaraderie, like we've both made the practical decision that it would be a waste of a night, and of all the alcohol we've consumed, not to at least attempt to have sex.

The first thing I notice in his room is the zombies: posters of zombies, zombie action figures and other zombie paraphernalia on every wall and shelf and surface. I don't ask about it. He doesn't want to have sex on the bed, and pulls a mattress from against a wall on the other side of the room, saying something about preferring it. I realise instantly that he's renting a shared room, that the mattress is his bed and that his roommate is away, but I don't comment. A series of thoughts flash through my mind; I tell myself not to think less of him, then I tell myself it's bad to have even considered thinking less of him for renting a shared room in the first place, when I live at home with my parents.

Moments later, lying on the mattress, I look up and notice a baseball bat fixed to the wall, still in its box and sealed under transparent plastic. The bat is a replica, wrapped in plastic imitation barbed wire and smeared with red blotches of paint made to look like blood. It's a prop from *The Walking Dead*, hung up next to a framed picture of Apocalypse Date #4 standing next to a man with long grey hair who, he tells me, later, is the show's special effects artist.

The following morning, I will find out that the bat has its own Wikipedia page; it's a symbol of violence, but also of undying matrimonial loyalty. Its name is Lucille, after the dead wife of the character who wields it. He keeps her spirit alive by talking to the bat, lovingly, as he uses it to bash other characters' heads in. You can buy replica Lucille bats online for €59.49; they're sold as 'The Walking Dead Lucille Bat – Take It Like a Champ Edition'.

At 2 a.m. I call a taxi, something I feel vaguely guilty spending money on, but I want to get home quickly. Outside it's freezing.

He walks me to the door, and it's only when we say goodbye, and he turns to go back inside, that I realise he's wearing a *Rick and Morty* T-shirt.

Ads for dating apps almost always mention new beginnings, but very rarely claim that a relationship will last. They're perfect for a generation denied stability – homeownership, job security and the chance to grow up. None of the men I met seemed happy to be in Dublin, and few of them planned on staying here much longer. Bumble introduced me to a succession of nihilists, due to their circumstances rather than by choice. Why call them Apocalypse Dates? Because they live under the same cloud as me, the same cancelled future. Because it felt like I met them at the ends of the earth.

Perhaps this was what I was waiting for: not the end of the world, but the end of a phase of life. In ways it was also an end of hope; when I open Bumble now, I am confronted with a list of disposable men. I hate myself for seeing them this way, but I know I'm disposable to them too.

Abandoned conversations fill the screen, people who I 'connected' with, in the sense that we used the same app, at the same time, and saw and liked each other's pictures. But I didn't make the first move, or he didn't reply, and if you wait a day then the match expires, and you lose the chance to talk to each other. There were simply too many, and we were too busy, waiting for someone better to come along.

In my case, someone better lives in my inbox. My ex, and my feelings for him, have survived hundreds of swipes and tens of Bumble matches. I am hopeful enough to keep writing to him, and pessimistic enough not to tell him I'm in love with him.

Dating was supposed to bring possibility into my life, to make me feel like I could have a future. Instead it feels like I'm running out of options; like the men I meet, and even the men I swipe through, all represent 'types' with whom I am incompatible.

Perhaps I am the target market for a lifetime Bumble membership, the customer they mark out as undateable. The world is shutting down rather than expanding, driving me to retreat, and to hide behind a screen, as coward and as judge.

I am hiding there still, alone, a permanent lurker, screening and sorting through the last men on earth.

Men Explain the Apocalypse To Me 2: Last Days

LOOKING BACK, IT FEELS LIKE my apocalypse dates were doomed from the start, because of the medium through which I met them. There was a cynicism to our meetings, a distance, like we never really left the internet, even in real life. I looked for reality in the wrong places; I stepped into an apocalypse culture.

The dates left me with questions about relationships, about myself, and about the dystopia around us. What led us to the point where any of this is normal? Is there no alternative? Perhaps the apocalypse is everywhere now; I blame dating apps for profiting from it, but not for creating it.

I've lived at home with my parents, on and off, for the better part of my twenties. I've moved away sometimes, to London, Edinburgh and Barcelona, but I always end up coming back. Over the years this has been a source of anxiety and guilt, and of fear that my future was cancelled before it ever properly began.

Apparently living at home is common in Dublin. I know this because of its impact on my sex life. I have tiptoed through the halls of houses in the suburbs late at night, to engage in near-silent sex with others who are also still living at home. I have accidentally met their parents. I've brought dates back to my parents' house too, but only very occasionally.

Things could be worse. As I write this, over 10,000 people in Ireland are homeless, and over 30,000 are on a waiting list for public housing, due to the failure of politicians to address the issue at present and in the past.[1] It can also be seen as a result, I believe, of the rise of the tech sector in the city where I was born.

In January 2003, a delegation of Google executives arrived in Dublin and were taken on a tour of the city's business parks. They decided to rent 60,000 square feet in a building yet to be completed, on Barrow Street in Dublin 4. Google chose Dublin over Switzerland as a location for its EMEA office, a formative moment for Irish tech.

Before that visit from Google, and the construction of the Barrow Street Googleplex, Ireland was already home to IBM, Microsoft, Oracle, Dell, Intel and others. After Google, however, the number of tech multinationals multiplied; Amazon, Facebook, Salesforce, eBay, PayPal, Etsy, Dropbox, Twitter and more arrived, filling the city with sprawling, primary-coloured campuses and transforming Dublin into an outpost of Silicon Valley. Today the investment made by US companies into Ireland is greater than that of Russia, China, Brazil and India combined, although many of the tech multinationals are known to pay less than the standard corporate tax rate, a low 12 per cent.

As with cities like San Francisco, it seems impossible for the arrival of big tech and the simultaneous sky-rocketing of Dublin rents to not be connected. In 2014, six years after the company was founded, 'sharing economy' rental platform Airbnb arrived in Dublin, leasing a 40,000 square foot, renovated nineteenth-century warehouse in Hanover Quay as their office and recruiting for 200 new positions. In 2018, the same building was occupied by protestors from the campaign group Take Back the City, who issued a statement mentioning the problem of evictions so that landlords could rent their homes on Airbnb instead of taking in long-term tenants, and highlighting the 3,165 entire properties advertised on Airbnb in Dublin at the time, in contrast with the 1,329 available on other sites for long-term rent.[2]

Around the same time that Airbnb were expanding their Irish workforce, I began to hear about 'Airbnb slums' – properties crammed with beds, and available for low-cost, short-term rentals – appearing around Dublin. I began to notice a certain kind of ad appearing on sites like MyHome and Rent.ie, alongside the €1,000-per-month studios, the Monday-to-Friday student rentals and the dubious backyard sheds rented as self-contained apartments. Now there were also shared rooms, costing around €300 per month, but frequently more, with several beds and mattresses crammed into one room. They were located close to the city centre, and leased to new arrivals who couldn't afford a more private place to live. I found these rental ads especially disquieting; the price of a shared room was the same as what I'd paid for a single one, five or six years earlier.

I also noticed a new kind of work culture in Dublin. Following the example of Google and Facebook, tech companies began to recruit with batteries of standardised tests and multiple rounds of interviews. Acquaintances who were accepted into these companies tended to disappear, like their social lives had abruptly ended. It was as though they'd given their whole lives to their work, through zero-hour contracts and unerring commitment of the spirit – the 'do what you love' approach, even if you clearly don't love what you're doing (nor do your bosses care, anyway). Companies demanded more, and offered less to those they employed. They discouraged unions and collective action, required longer and later hours, and were able to fire people for lack of enthusiasm, or lack of morale, demanding employees sign non-disclosure agreements upon dismissal.

Other companies in Dublin were inspired by the tech

industry's example. Throughout my twenties I heard more and more employment horror stories; well-known organisations that ran on unpaid intern labour, creatives whose work was ripped off, who were overworked and emotionally manipulated, treated as disposable, and asked to work 'for exposure' instead of being paid. Employers didn't seem to care that Dublin is tiny, and that word would get around if they treated people badly. With a surplus of young people accessible through social media, they could easily, guiltlessly find another mark who was willing to work for free.

Rents continued to rise. I worked, but still I could not afford them. This in itself felt apocalyptic, like my ability to plan for the future had been taken away. For years I wrote, and that writing helped me to find meaning in my life, but it fell short of making me feel like an adult, or like any less of a loser.

Over the years I went through phases of optimism, but mostly long, unbroken periods of pessimism. Where it hit me hardest was in my attempts at relationships, which seemed almost like they were long-distance, even when they were with people in the same city. It meant waiting for someone's parents to go out in order to have sex. Sometimes it meant booking cheap hotel rooms. Sex became rare and anticipated, and this created pressure for that sex to be good. If the guy I was seeing did have somewhere of his own to live, it meant I'd always have to stay over, and this created a strange imbalance of power, like I was delivering myself to his door on demand.

In every case, it felt like a mockery of adulthood. My actions were confused, muddled by a quiet desperation. Over the years, I could never be sure if I was dating these guys because I wanted to, or because I knew that relationships were what independent, grown-up people had.

*

In 2018, a study found that 'women survive severe famines and epidemics better than men'. The researchers surveyed historical conditions carrying 'high, perhaps extreme levels of mortality risk', including the French cholera epidemic of 1832, the Irish Famine (1845–52), nineteenth-century slavery in Trinidad and the migration of freed slaves to Liberia (1820–43), and found that women had a 'survival advantage' and led longer lives.[3] The researchers do not discuss the reasons why this is the case, but in an interview one of the authors of the study, Virginia Zarulli, proposed that it might be due to a mixture of behavioural and hormonal factors. Oestrogen protects the immune system, while testosterone can suppress it. Testosterone can also encourage reckless behaviour. Zarulli told Live Science 'The female survival advantage has deep biological roots, but the role of culture, society and behaviour is very important as well.'[4]

Would I survive in an 'extreme' environment, one that's physically extreme, rather than just mentally so, like the internet? Would I survive an apocalypse? I don't really know. I'm 5 foot 2 and not especially strong. My strategy, I think, would be one of self-preservation rather than attack; I would climb a tree to hide at night, from men and marauding wild animals. I like to think Bumble, among other things, has taught me the social side of survivalism; I'm good at flattering strangers in order to get them to like me, and I'm good at trying to forget them when the relationship has run its course.

It's likely that there will be more post-apocalyptic women than men, although doomsday prepping, the largely American, largely online movement dedicated to apocalyptic anticipation, strikes me as a distinctly masculine practice. Post-Y2K and post-9/11, an industry has flourished around cataclysmic forecast, fuelling sales of guns, military-inspired survival supplies, dehydrated food contained in ration buckets, 'tactical' clothing, 'bug out bags' (portable apocalypse survival kits) and combination weapon-tools like the Crovel Extreme II, an American-made shovel with a built-in saw-edge, crowbar, hammer, axe and, should you need it, bottle opener, allowing its user to ward off attackers, saw apart prey, bury the pieces and celebrate with a bottle of beer, all while using the same tool.[5]

Popular figureheads associated with the movement confirm the archetype of a white, conservative man; as I write this I scroll through a webpage, Infowarsshop.com, where the media personality Alex Jones has curated an apocalyptic grocery list.

Jones is a public-access TV host turned far-right conspiracy theorist, known for calling the Sandy Hook school shooting a 'false flag' operation, and for implying that 'chemicals in the water' are turning frogs gay. His online shop sells prepper necessities; my favourite, the 'Tactical Bath', is no longer available, but it gained online infamy back in 2017. It was a moist towelette, like a bigger version of the sachets handed out after you eat chicken wings, packaged in black and gold and sold as a body-safe 'no rinse system'. The Tactical Bath was Jones's vision for post-apocalyptic personal hygiene; it is 'Pentagon technology now available to the public', the website claimed, sold for $9.95 and suitable for sensitive skin around 'the perineal area' and mucous membranes. In the top right-hand corner of the webpage is a picture of Jones himself, looking earnest and digitally filtered, along with the quote, 'Thank you for supporting the infowar'. His message draws commerce into an ongoing struggle to overcome what Jones sees as a global conspiracy, suppressing truths about 'false flag' operations, covert weather control programmes, human–animal hybrids, and an apocalyptic plan by 'globalist elites' to destroy America.

The Tactical Bath is notorious among preppers and those who mock them online. It was met with disbelief on social media, and mocked on the TV show *Last Week Tonight* by its presenter, John Oliver, who accused Jones of selling 'sloppy wet rags for your taint'.[6] It gave form to prepping's forlorn commercial ambitions, the drive to conquer the apocalypse – and, it is implied, death – with the same disposable, ultimately useless consumerism that will likely have caused it.

I find, in the Tactical Bath, and in the digitised smoothness of Alex Jones's face, a kind of lurid, fastidious camp. I am moved by his sanctimonious volatility, his need to be young on the internet for ever, coupled with a barely concealed longing for the end of civilisation. There's something about this bland, skin-coloured picture of Jones, and the idea that anyone who isn't joking would buy a Tactical Bath, that makes me feel intensely sad. I can barely keep myself from crying while looking at it on the screen (this is likely the result of my female, immunity-boosting hormones). It brings home to me the idea that these longed-for end times will be a kind of pathetic magic sprinkled over mundanity, transforming everyday objects, and people, and assigning them new purpose. It *will* be a transformation, but one of shit into more mundane shit, kipple into kipple, a monotonous echo of what made us apocalyptic in the first place.

My belief, it seems, goes against popular apocalyptic thinking, because it is their sense of purpose that doomsday preppers prize most of all; the belief that when the time comes, each of them will be proven rare, rather than disposable. As the last men left on earth, everything they do will carry weight, and importance, and an air of destiny. Prepper culture, with its drills, its emergency rations and its 'tactical' consumer products, is heavily inspired by the military and by action film depictions of masculinity. There's the barely latent implication that, once shit hits the fan ('SHTF'), the apocalypse will become your licence to kill: zombies or not, you'll be fighting off opponents with your Crovel Extreme II, removing heads and destroying brains, and protecting your family with a shotgun (it's interesting how the left fetishises a similar cleaning of the slate in the form of revolution – the difference is that at least there are plans for the utopia to come, beyond stockpiling buckets of maize and jugs of filtered water).

However, there is a tension within the prepper community between conspicuous preparedness and another survival strategy: deliberate anonymity. 'Grey man theory' hypothesises that in a kill-or-be-killed environment, the man who looks like everyone else can more easily keep out of harm's way. In the great disposal of the civilised world, you can cheat death, by making yourself seem disposable. The 'grey man' approach goes against other visions of apocalyptic masculinity because it involves blending in as much as possible; no visible weapons or survival tools, only 'pedestrian' clothing, and no high-tech accessories that will attract attention or thieves. Would-be grey men are advised to go along with the crowd, and to adopt a friendly, innocuous manner. The kind of manner, it strikes me, that women have always employed when negotiating with men.

In December 2012, the year and month of the forecasted Mayan apocalypse, Durex released an apocalyptic marketing campaign, featuring slogans like 'Go out with a bang', 'See it right through to the messy end', and, 'The end is coming'. Why anyone would bother with condoms, in the event of an apocalypse, was left unexplained.[7]

That same year, an LA-based porn studio called Pink Visual announced plans for a luxury bomb shelter capable of holding up to 1,500 people.[8] The studio planned to publish an open call for bunker inhabitants, albeit with selection criteria, and a bias towards performers, business partners and fans. Coverage in Russia Today describes it as 'the Bio-Dome but with blow jobs', noting that it will include 'multiple bars with premium liquor, an operational microbrewery, stripper poles, topless dance stages and, of course, "a sophisticated content production studio"'.

The apocalypse has everything to do with sex, and it has everything to do with money. On one level, it's clear that we have sex to remind ourselves we're alive; in evolutionary terms, it can be a reminder that we have a future.

In this sense, sex is the antithesis to an apocalypse, but the two keep getting put together. This applies especially to casual sex; it's as though it is easier to imagine the end of the world than to imagine the end of monogamy. During my Bumble dates, I kept wondering why every man I met wanted to talk about the apocalypse, and whether this connection was conscious, or unconsciously made. I thought about how all of them worked, or wanted to work, in tech, where our secrets and friendships and behaviours – the essence of what makes us human – get processed and sold as ad data in bulk. Proof of technology's effectiveness – its inevitability, even – was that we had 'matched' through a mobile app, and were now prepared to have sex with each other.

In August 2015, *Vanity Fair* published a piece by Nancy Jo Sales titled 'Tinder and the Dawn of the "Dating Apocalypse"'. It announces an end to time-honoured rituals of courtship, portraying the app as a marketplace for casual sex-on-demand. Throughout the piece, Sales follows young people in Manhattan's financial district as they swipe through apparently endless partners. She frames the arrival of swipe-based dating apps as an evolutionary event, listing it alongside the melting of the polar ice caps, as a force speeding us along the path to annihilation: 'Hookup culture, which has been percolating for about a hundred years, has collided with dating apps, which have acted like a wayward meteor on the now dinosaur-like rituals of courtship.'

Sales goes on to quote a research scientist, who warns that 'we are in uncharted territory', and a twenty-something investment banker named Dan who says, 'It's like ordering Seamless [a food-delivery app]. But you're ordering a person.'

This 'dating apocalypse' marks a beginning as well as an end;

it's the ultimate form of what tech calls 'disruption', destroying the old and making way for the new. But columnists and pundits and even sociological studies lamenting the rise of casual sex have been around far longer than dating apps. They go back several decades, perhaps even further back than the 'Happy Families Planning Service', the first computer dating program, created at Stanford in 1959.

The truth about casual sex and technology is surely more complicated than the phrase 'dating apocalypse' implies. Data shows that people are using apps to find long-term partners as well as one-night stands, and a 2017 survey of over 14,000 newly-weds and newly engaged people found that 19 per cent of brides said they'd met their partner online, with dating apps ranking above meeting through friends, at college and at work. Marriages between people who met online were also less likely to end within the first year, and the couples expressed more 'marital satisfaction' than those who met in other ways.[9]

And yet, it seems obvious to me that dating apps have changed dating behaviour. It also seems obvious to me that these apps, which are predicated on data and probability, and which demand a similarly cold, statistical aptitude and a readiness to take just-suggestive-enough selfies in order to succeed at them, have created a certain mindset in their users. I don't think I'm alone when I say that prolonged exposure to Tinder, Bumble and their ilk leave me feeling vaguely like a sociopath. They leave me feeling like someone scoping out victims to lure into a car, or like the alien creature in *Under the Skin* that masquerades as a woman, the one who kerb-crawls through Glasgow in search of innocuous lads to abduct and process into food.

I came to resent Bumble because of its neediness, which almost rivalled my own. I noticed its transparent attempts to manipulate my behaviour, and how occasionally, on weekends, I'd get push notifications telling me 'Bumble is on fire! You're 3.2x more likely to get a match.' My bio, too, seemed to be a constant source of concern to the people at Bumble. They'd issue reminders laden with emoji and exclamation marks: 'Our data shows that a short & punchy bio increases matches!' My bio had worked perfectly well for me so far. It was a single line from *It's Always Sunny in Philadelphia*, spoken by Danny DeVito; 'Can I offer you a nice egg in this trying time?'

Something else was getting to me about Bumble: the combination of emotional labour and cold, hard-edged reserve demanded of its users in order to make it work. At times Bumble felt more like a recruitment app than one meant for dating. I came across some truly awful profiles. One young professional had made his profile a picture of himself meeting the Taoiseach, Leo Varadkar. Another guy's profile picture was a close-up of his own face, so close that the image was only a fleshy, diffused blur with two faded eyes in the centre of the screen. There were the prescriptive types, who wrote that they never drank and never smoked, and that their matches couldn't either ('not into bars and drunk hookups; I have too much self-respect') and the guy who would date only Muslims, Christians and Jews. There were men who used business headshots in their profiles, and one whose bio spoke of his longing to escape the office, and his plans to hike the Appalachian Trail. I found him inexplicably tragic, like some hapless businessman from a George Saunders story. There were also many otherwise forgettable men who used their bios to complain about the women they encountered: 'Big no's to Snapchat filters, duckface, loops and excessive aerial-angled selfies. I'll more than likely dislike you so save yourself the trouble.'

Bumble was making me hate men, even as it kept me searching for more. It was making me opportunistic; if apps had somehow 'gamified' dating, then sex, too, had become a point-scoring game. Eventually I admitted to myself that I wasn't even looking for a relationship on Bumble; I wanted someone who wasn't there. I continued to write emails to my ex, hundreds every week, which I came to consider a kind of relationship. I told myself that this was a modern condition, the compartmentalisation of love and one's sex life.

I planned to become a Bumble queen, cold and resilient, a trafficker of mediated affections. For a time I got high on it. Men were disposable. Men were a crowd to be sorted into haves and have-nots. Men were a game. Men were the little primary-coloured ghosts in Pac-Man, running at me, and I dodged them or ate them alive. The app brought out something unnatural in me, but it also made me feel like I was staying afloat. I congratulated myself for becoming this loathed thing, the millennial nightmare, feckless, voracious, craving intimacy yet emotionally unavailable.

It was only afterwards that I began to wonder where this cynicism came from. Perhaps it was in my nature, or perhaps it was it what the platform, Bumble, demanded of its users.

Many of us have jobs that are waiting to be filled by robots. Human labour, in this era, just happens to be cheaper, but soon AI will be more affordable to the average business. The same logic applies to our love lives. When the Basic Pleasure Models come in we'll form an orderly line to buy them, but until then, we'll use the closest thing to an automated lover – the automated distribution of lovers – to stave off our loneliness.

Bumble has over 100 million users at the time of writing.[10] Seventy-two per cent of them are under the age of thirty-five. The amount of time these users spend on the app averages 100 minutes per day, making 2 million swipes per hour, and generating, for Bumble, a yearly revenue of $200 million. Among that number, over 10 per cent pay for a monthly subscription, Bumble Boost, which includes features like 'Beeline', a list of users who have already liked you, 'Rematch', for retaining matches that have already expired,

'Busy Bee', which allows for match extensions, and 'SuperSwipes', a way to demonstrate interest in someone, for £7.49 per week. Longer-term subscriptions are cheaper – three months costs £41.99, and for £79.99 you can subscribe to Bumble for a lifetime.

This model is common among dating apps, a field led by Tinder with its 'Tinder Gold' option, costing $14.98 per month. Gold was launched with comically hyperbolic, elitist language, describing it as a 'first-class swipe experience' and 'members-only service', offering 'our most exclusive features' (I've noticed that 'luxury' internet products like Tinder Gold never promise to decline processing their user's data, an option I suspect will one day be valuable, but which for now is probably not cost-effective to the owners of these apps).[11]

Dating apps would like their users to believe they're receiving an elite online experience when they upgrade, but really they are paying for some small portion of the omniscience and power retained by the people who run these apps, who maintain a God View over their users and can freely manipulate their behaviour. Dating app founders are like tattoo artists; every day they take actions that will potentially affect the rest of a stranger's life, but they carry this burden lightly. I've never seen a dating app founder admit that they deal in our civilisation's future, the preservation of a lonely species doomed to languish in front of screens.

Each year new terminology emerges to describe strange and inhumane behaviours common to dating apps. 'Ghosting', famously, means cutting off contact. 'Benching' is when you put someone aside, potentially for later, 'breadcrumbing' is when you continue to Like their social media posts, and 'zombie-ing' is when a seemingly dead match 'reanimates' and sends a 'Hi, how've you been?' People don't use these terms aloud very often, but I'm certain that they experience and practise them. Quietly, we have grown to treat others the way Silicon Valley treats us: as data subjects, as nodes in a system.

Tinder once ranked its users internally with an index called the Elo score, a ranking system more often used for judging ability in chess and other zero-sum games. On Tinder, it arranged users in terms of their desirability. Each time your profile was swiped, left or right, your ranking changed; the more desirable someone was judged to be, the more options they were given.

That same logic of competition, and of humanity made to fit within digitised systems, pervades the dating app user experience; at best, apps are 'gamified', addictive and fun, but at worst they make flirting into something grim and transactional, a process where chat-up lines are repeated again and again. Faces and personalities blur, and the belief manifests that attraction, and people themselves, can be 'hacked'. It's hardly coincidental that 'PUAs' (pick-up artists) have flourished in this landscape, selling courses, e-books and access to online communities that teach lonely people how to attract partners. Through dating apps, we are taught to flirt and to love like robots, then we are surprised when we inevitably feel nothing.

A number of commercially available dating app bots have sprung up in recent years, which promise to do the swiping and flirting for you. In 2014 a Java developer named James Befurt built a device consisting of a microcontroller, a piston, a stylus and a corresponding computer program, which could auto-like profiles on Tinder.[12] A spate of Tinder auto-like programs subsequently appeared and circulated online despite being banned from the App Store and prohibited by Tinder's terms of service. They joined a nascent cottage industry, encompassing dating coaches, articles suggesting 'Tinder closers' and 'Tinder hacks', and even functionalities built into the apps themselves to help users come up with the perfect chat-up line.[13]

There's a vaguely depressing video online which shows the Tinder swiping machine in action. It emits a low hissing noise as the stylus swipes, at a rate of roughly one profile per second.

It collects women who, somewhere on the other side of the app, might believe that guys on Tinder actually take the time to look at their profiles. The video is titled 'How an Engineer Uses Tinder', which reflects rather dubiously on engineers.[14]

Swiping machines, however, are only a small part of a broader problem. In online media, much is made of how unpleasant dating apps are for women, how they face chauvinist insults, unsolicited dick pics and messages from creepy, persistent strangers. All of this is true. In this sense, the problems women experience on dating apps mirror problems women experience on the internet at large. Dating apps and their issues, which can be summarised as 'the dick pic problem', speak to the brief yet complicated history of a male-dominated online sphere – one where, in the modern internet's early days, forum users were assumed to be male as a default. 'A/S/L' – 'age, sex, location' – was a common question, and if the user responded that they were female it could be followed by a demand for 'Tits or GTFO'. Another phrase, long fallen out of use, comes to mind here: 'There are no girls on the internet' – Rule 16 from 4chan's 'Rules of the Internet', created in the mid-2000s. This line is at once dogma and lament; it was assumed that anyone claiming to be female on popular channels, like Usenet and multiplayer games, was a man in disguise trying to trick you, but the rule also legitimised misogyny at such a level that women didn't always feel welcome or even safe using certain sites.

Dating apps are disproportionately weighted in favour of men in terms of their design – it seems obvious to me that the visual nature of apps like Tinder caters to stereotypes around male sexuality, and that the very idea of geolocated dating, which took off with Grindr before going mainstream and hetero with Tinder, flies in the face of typically female concerns about safety, privacy, vulnerability and stalking.

However, despite all these things, to be female on dating apps today is to find oneself in the rare position of having an inherent advantage. While apps are essentially designed by and for their use, men face a different sort of struggle, one that is every bit as dystopian, and late-capitalist, and dehumanising, and cruel. To be male on an app like Tinder or Bumble is to be one of many, almost infinite matches. It means acknowledging that you are predictable, mundane and utterly disposable.

Men outnumber women on location-based dating apps by almost two to one – a 2015 study indicated a 62 per cent male user base on Tinder worldwide,[15] while study of its UK users, from 2019, placed the number at 85 per cent. Roughly 1 billion Tinder swipes are made per day, but only 12 million of those turn into matches, and the number of women returning male attention is statistically low; while men swipe right (i.e. 'like') 46 per cent of the time, women only do so for 14 per cent. No wonder the majority of the auto-like programs, as well as the swiping Tinder bot, were built by male engineers.

Another, significantly more casual study (published online, without peer review) was conducted by an anonymous author under the name 'Worst-Online-Dater', and published to *Medium* in 2015.[16] It estimates that the bottom 80 per cent of male Tinder users are battling it out over the bottom 22 per cent of female users, and that 78 per cent of the women on Tinder are competing over the top 20 per cent of men. Men 'like' women 6.2 times more often than women 'like' men; the odds are against them, even if they use a Tinder bot to automatically swipe right on every woman.

The author then uses the Gini coefficient, an index used more often to estimate wealth inequality, to calculate the inequality level of Tinder, and finds that for male users, the app is less equal than 95.1 per cent of the countries in the world – the only countries ranking statistically below it are Angola, Haiti, Botswana, Namibia, Comoros, South Africa, Equatorial Guinea and Seychelles.

This means that an average-looking man will be liked by below 1 per cent of the women using the app, acquiring one 'like' for every 115 women. Viewed this way, the odds of finding someone who will sleep with you are near non-existent, the odds of falling in love even less so.

When she left Tinder and founded Bumble, Whitney Wolfe Herd filed a lawsuit against her former employers, alleging 'atrocious sexual harassment ... representing the worst of the misogynist, alpha-male stereotype too often associated with technology start-ups.'[17] The case was eventually settled; *Forbes* reported that Wolfe Herd received over $1 million in Tinder stock.[18]

Not every tech workplace is toxic, or misogynist, but they are often predominantly male spaces. Among dating apps, almost all of the best known (Tinder, Match, Badoo, Hinge, OkCupid, Happn, Plenty of Fish, eHarmony) have male founders. This brings the dissatisfaction of their male users into a sharper focus; dating apps, it strikes me, are a male dystopia built by male engineers.

There's a website that asks 'How Many Apocalypses Have I Survived?'[19] You type in the year of your birth to find out the answer. My result, calculated in early 2019, is forty-four apocalypses, starting in 1989 with a prediction of former NASA engineer Edgar C. Whisenant, who continued to make predictions even after the failure of those made in his book titled *88 Reasons Why the Rapture Will Be in 1988*. My apocalypse list progresses to a prediction for 1991, made by Nation of Islam leader Louis Farrakhan, that a 'War of Armageddon' was beginning, a prediction by self-professed telepath Sheldan Nidle that in 1996 alien spaceships and angels would descend on Earth, the mass suicide of the Heaven's Gate cult the following year, the Y2K bug, the Nuwaubian Nation's 'star holocaust' of 2000, and the 2010 apocalypse augured by the Hermetic Order of the Golden Dawn.

In June 2019, the year I write this, I hope to survive another apocalypse. This time it's the return of Jesus, predicted by Ronald Weinland, a convicted tax evader and leader of the Church of God Preparing for the Kingdom of God, an 'apocalypticist splinter sect', as it is described on Wikipedia, formed in the late 1980s. If you are reading this after that date, then congratulations on your – our – survival.

Sometimes I think about how convenient it would be for the world to shut down, for everything around me to be suddenly switched off. It would neatly mark the end of all our messy centuries. It would prove that few – maybe none – of our lives mattered, for once and all. I wouldn't have any more deadlines, I'd not have to pay off my student loan, and suddenly all the things that have tormented me – the drive to achieve, the need to finish writing this essay, and the long, fulfilling relationship I wish I had – would fall into a pit along with our flailing, combustible bodies.

Then I remember that this is what nihilism is. I don't think of Tinder, or Bumble, for that matter, as signs of a 'dating apocalypse'. They're symptoms of a modern problem, that of loneliness, for which they are also a temporary cure.

There are other, more apocalyptic symptoms out there, like a creeping animosity between people, which apps geared towards romance and communication claim to remedy but which they more often advance. There is acrimony between genders, between digitally imposed filter bubbles, and between social groupings that don't even need to exist. They're built on fictions, the kind that social media platforms encourage us to tell about ourselves. A sense of competition persists, like a mutated version of the competition that capitalism demands of us.

In his book *Heroes* (2015), the philosopher Franco 'Bifo' Berardi addresses 'the establishment of a kingdom of nihilism and the suicidal drive that is permeating contemporary culture'. He discusses an 'annihilating nihilism' within modern capitalism, one which destroys moral and economic values in order to affirm the abstract force of money. Work is a 'network-machine', one that consumes time and individual identities, until its toxicity filters into everyday life. Berardi then explores the connections between this particular kind of predatory capitalism and mental health, describing the psychology of Eric Harris and Dylan Klebold, the perpetrators of the Columbine school shooting, as a 'suicidal form of the neoliberal will to win'.

My lifetime, and in particular the past two decades, have been marked by a dramatic increase in school shootings, mass shootings and other, very public acts of mass murder and suicide. Most of them take place in America, with perpetrators that are almost universally male, and below a certain age. With a growing frequency, they are almost always active on social media, which serves to document their radicalisation, their plans, and their final, fatal rampages. In May 2018, after a van rampage in Toronto by a 25-year-old software developer called Alek Minassian which killed ten and injured sixteen, the Southern Poverty Law Centre added 'male supremacy' to the list of hate groups they track.[20] Included, in a subcategory, were incels, an online community of 'involuntarily celibate' men that congregates on Reddit, 4chan and other sites with names like 'sluthate.com'.

The list of attacks associated with the incel subculture is prodigious. It includes at least eleven known examples, beginning in 2009, where the killer was either actively part of the community or embraced by it after his death. The incel-associated murderers, including Nikolas Cruz and Elliot Rodger, the self-proclaimed 'supreme gentleman', share common traits. They experienced problems with women, failing as other young men around them were able to succeed. Each mass shooting is also a suicide mission: none of these men could see a future for themselves, so they decided to end the futures of others, too.

For all its antisocial, occasionally antinatal beliefs, the incel subculture has attracted tens of thousands of adherents (it's difficult to estimate, as its communities keep getting quarantined or shut down) to a stark worldview where sex, capitalism and self-worth are entwined. The reasons incels provide for opting out, and for hating the 'Stacys' and 'Chads' (good-looking, sexually fulfilled or promiscuous people) who dominate the sexual marketplace are vaguely rooted in evolutionary biology, and sprinkled with ideas and terminology imported from quack disciplines like phrenology, 'race realism' and bro science elements of fitness culture. They believe rigidly in a system of 'mate value'; that women are drawn to high-ranking, 'Chad'-type men, leaving others to inceldom. The value of women, in this sexual marketplace, declines with ongoing promiscuity, until they pass their sell-by date at thirty and are rendered a 'roastie', i.e. their vulvas somehow grow to resemble roast beef.

Along with its misogyny, there are other elements to the incel subculture I find disturbing, and believe are worth taking seriously as a sign of something being very wrong on the internet. The first is their implicit belief that women can save them, and that sex, and a faithful monogamous relationship, will be enough to bring meaning to their life. The incels miss a trick in assuming that women are content with there being a 'sexual marketplace' in the first place, when many of us are clearly miserable, and tired of living on these terms.

The other thing I find a little disturbing, and fascinating, is the solution they propose to their problems. In 2013, a blog with the URL 'governmentsgetgirlfriends.wordpress.com' published a post titled 'Programs for treating love-shyness'. It proposed a girlfriend distribution programme where women are offered money to go on blind dates with 'incel' men:

> Every woman would have a limit of thirty dates. If she doesn't find a suitable partner during those thirty days she will be fired

> to prevent scammers – however, she would be paid the full sum, as would a woman who finds a partner during one of these thirty dates ... Using this program, many involuntary celibate men would get their first date or improve their chances of finding a partner.[21]

Dating apps earn money from lonely men, rather than the women who go on dates with them, but it should be obvious that the incels and the creators of dating apps occupy two sides of the same coin. They inhabit the same neoliberal dating culture where the brute categorisation of humans is offered as entertainment, where people are disposable, where virtually all sexual attraction is looks-based, and where, every time we 'interact' and 'empower' ourselves to make the first move, we create profits for a small number of entrepreneurs in offices in Silicon Valley.

This does not justify the actions of incel killers, or even the broader community's casual misogyny, but it does lend them context. A culture in which all things can apparently be 'hacked', yet some men can't 'hack' attraction, will inevitably lead to disappointment. As one community member writes, on the subreddit 'Braincels': 'Too late for family values, too early for AI robo wives, just in time for hypergamy. What a fucking straw we drew.'

In 2011 came the 'Forever Alone Involuntary Flashmob', where a group of lonely men were duped with fake dating site profiles into congregating on a Friday night in Times Square, New York. They'd been led there by a campaign organised in secret by 4chan users, posing as women on dating sites, who watched what ensued through a public webcam. Brayden Olson, writing for *Vice*, described the scene:

> After peering at it for about five minutes, I realised that I was sat by myself in the dark at home at 1 a.m. on a Friday night, pitying at

> a bunch of guys for supposedly being 'forever alone'. The sudden sense of irony was pretty crushing.[22]

Today the 'forever alone' experience lives on in bots posing as sexually voracious women on dating apps, and in the meme, shared among incels, where men reply to advances made by these bots, posing as attractive young women, with the line 'just give me the virus link'. Tinder bots are capable of holding automated conversations for just long enough to fool lonely men into being scammed. Their profile pictures invariably depict a young, overtly sexy woman whose comments are forward and flirtatious, promising webcam action, or further conversation, if the user is willing to click on a link and go 'off-app'. This leads to ransomware, or spam, or, if they're lucky, an innocuous advert for porn or escort services.

Bots are the tax for male desperation, a punishment for loneliness. They played a significant role in another online controversy in the 2010s, the leak of data from Ashley Madison, an online matchmaking service for people looking to cheat on their partners. In July 2015, a collective called the Impact Team stole over 25 gigabytes of data from the site's parent company, Avid Life Media, including user details, pictures and, allegedly, the source code for their products. The hackers issued a demand: if Ashley Madison, and its sister company, the 'premium dating' site Established Men, were not shut down immediately, they would release the identities of their users to the public.

Just under a week later, the information was posted as a 10 gigabyte compressed archive, linked on a page on the dark web. The file was cryptographically signed with a PGP key, with a message from Impact Team:

> Time's Up! Avid Life Media has failed to take down Ashley Madison and Established Men. We have explained the fraud,

> deceit, and stupidity of ALM and their members. Now everyone gets to see their data. Find someone you know in here? Keep in mind the site is a scam with thousands of fake female profiles. See Ashley Madison fake profile lawsuit: 90–95 per cent of actual users are male. Chances are your man signed up to the world's biggest affair site, but never had one. He just tried to. If that distinction matters.[23]

Among dating and hook-up websites, Ashley Madison stands out (it's still online, somehow, and functional, at the time of writing) for its pricing structure, a points-based, casino-style system in which male users buy bundles of credits in order to send and read messages received.[24] Women are allowed to use the service for free, but men are charged $59 for the 'basic package' of fifty credits, with five credits charged per message, and an additional five for messages marked 'priority'. Messages are, in fact, marked 'priority' by default, and users are automatically charged $79.99 for a new package of credits once the package they start with runs out. Credits are also charged for time – thirty minutes of chat costs thirty credits – and there is a $19.99 activation fee simply for downloading the mobile app. In an interview with *Motherboard* conducted over encrypted email, Impact Team accused Avid Media of making over $100,000,000 per year in fraudulent charges.[25]

Men can burn through their initial Ashley Madison credits within minutes of opening an account, but what made their situation even more dismal was that most of the 'women' they contacted were actually bots. Annalee Newitz, editor-in-chief of Gizmodo, analysed the data dump and found evidence of over 70,000 female chatbots designed to occupy Ashley Madison's male users, an 'army of fembots' employed in a bid to 'create the illusion of a vast playland of available women.'[26] Randomised 'engagers' and 'hosts' inhabited fake profiles called 'Angels', using profile photos taken from untended Flickr profiles or old, dormant Ashley Madison profiles. Newitz even found sixty-nine instances of bots talking to each other, as well as human users, by way of a glitch. There was evidence of men being passed to 'affiliates' after chatting with the bots for a while, likely escorts, and a proposal written by one Ashley Madison engineer to bring in a system of 'FemaleValue' and 'MaleProfit', in which human, female users would be paid for engaging with men. It reads like a parody of the 'GovernmentsGetGirlfriends' call for a kind of sexual Marxism.

In the wake of the Ashley Madison hack families were torn apart, marriages ended, and a number of suicides were linked to the release of user details. The hack was credited to idealism, to Impact Team's disgust at Avid Media's exploitation of their users, but it's hard to know whether to believe them; was this a strike against Silicon Valley cynicism, or simply more hacking, and humiliation, for kicks?

On Reddit I've seen commenters liken Tinder to a form of eugenics. Others, I've noticed, tend to criticise it for an overabundance of choice; they lose focus, and fail to settle when the right match comes their way. Personally I think that, much like Skype sex, it's easy to feel like you're engaging in a romance with your machine, rather than with a person, while using dating apps. Between their push notifications, their drive to keep you sorting through people, contributing data and *interacting*, they train us to be technocapitalism's perfect citizens, the ideal 'data subjects'.

I hope that our data will not be in vain. I hope that in the near future, when entire generations have gone without long-term relationships, the data generated by lonely people on dating apps will at least be put towards the design of sex bots, so that we can live out the remainder of our days with our Basic Pleasure Models in peace. Perhaps AI will even create the sexual Marxism the incels are asking for. If not, at least there will be DeepFakes, and VR porn, and ASMR to keep everyone occupied.

Lately it feels like I'm frozen in 'tonight', in an eternal present. For almost a decade, on and off, I've been living at my parents' home in Dublin, a city I was born in but still cannot afford. Nothing seems to change, and this gives me anxiety, a dull sense of doom which psychiatrists have called catastrophising. Precarity, depression and uncertainty have eroded my ability to imagine the future. A fog is draped over the present and the past, until all I have is the moment, *tonight*.

In the gloom of the eternal tonight, the clubs I spent my twenties dancing in close and never reopen. Friends move away, university degrees are proven worthless, and I find myself drifting, occasionally achieving the things I dreamed of, but feeling very little when I do. Studies have shown that depression shrinks the hippocampus, the organ in the brain responsible for memory as well as future thinking.[27] The arguments theorists like Mark Fisher and Franco 'Bifo' Berardi make, about the slow cancellation of the future, turn out to have links with neuroscience.

Every relationship is a promise of a future. Even a one-night stand, for the right kind of misguided romantic, can serve as such. What dating apps do is traffic in futures, enchanting us with possibilities only to rush us along to the next match, the next interaction, which earn them further profit as data. These apps harness technocapitalism at its most invasive, its most urgent and persistent, while concealing from the user their tendency to treat them as disposable. I went home with men, and all they wanted was to imagine the apocalypse. A quick, fiery end occupied the space where adulthood should be.

After I quit Bumble, I sent the app a subject access request asking for the data they'd retained on me. Bumble HQ replied with a message asking me to fill out a form and attach scans of two identification documents, a bank statement and my passport, in order to claim it back. They never asked for this information when I signed up.

Three weeks later, after a few emails to remind them, Bumble sent me a link and a password to download my data. I received files containing messages I'd sent, including only my own side of the conversation with gaps where the men had been, now absent data-ghosts. The conversations were short, and grouped together in one document. Then there was a PDF, headed 'FOR INTELLIGENCE PURPOSES ONLY', which listed the bare bones of my identity:

Language: English
Tattoo: no
Piercing: no
Profession: writer
Company name: somewhere
Lifestyle dating intentions: don't know
Lifestyle exercise: sometimes
Lifestyle family plan: someday
Lifestyle religion: agnostic

It's jarring to see oneself as data, the internet's equivalent to drawing blood. It's the digital abject, the point where you realise that you are what the internet is made of. It feels so generic as to be inaccurate; I'm not sure if I'd call myself agnostic, it's not the kind of thing I can tick a box in order to explain. My 'lifestyle family plan' is probably the most nebulous detail of all. I don't know if I even have a plan, or a coherent 'lifestyle' to speak of.

Maybe I do have a plan, but it's conditional on one person. I keep writing emails to my ex, searching for a connection, and time, which I believe is the test of love's intensity. I'm not looking for an apocalypse; I'm looking for a future instead. Even if he won't give me these things, I'll be content with emails.

Lately my life feels like it's over, but at the same time that life has become endless. Endless youth in an endless present. Endless encounters, but none of them lasting. Endless possibility and endless anticipation. Endless pages of search results. I will never run out of internet in my lifetime, and after my death my life will continue as data on a server, turning a profit for someone I've never met.

The way we speak of it, an apocalypse is the same thing as eternity; it's a pledge, a declaration of love, a curse, an impossibility, an act of hyperbole that reveals more about the present than the future. We fantasise that our lives will be cut short, as will everybody else's, because it's the easiest way out of the mess we're in.

Tamagotchi Girls

MY FIRST EMAIL TO YOU is dated 28 March 2015. Since then, we've exchanged 2,347 more. In the beginning, threads would drop off after ten or fifteen messages. Then it was twenty, seventy, ninety; layer upon layer of text, links, pictures, in-jokes and, for me at least, unspoken desperation.[1]

I carried your ghost on a screen in my pocket. I think I might even have summoned your tulpa. You existed to me as words, alerts and unread messages, hints and possibilities, and occasional heartbreaking meetings in real life, where I could never bring myself to tell you how I felt.

It wasn't clear where we were going, but I'd write to you with any excuse. We only dated for three months, but we stayed in contact for three years afterwards. During that time I retreated from the world – I broke down, went into treatment, then I started writing and recovering. I became suspicious of technology, but I couldn't detach from it, not when it was my only means of talking to you.

I found myself trapped in the screen's parameters. I couldn't make the first move; I lacked confidence, even when I was able to hide behind a laptop. But I always wrote back eventually, fuelled by the things I couldn't say: that I was thinking about you, constantly; that I loved you; that all this love had appeared, and I didn't know what to do with it.

Over those three years I tried, and failed, to have relationships with other people, and your messages continued. I'd read one and feel myself swallowed, the world outside the screen falling away. They offered fragile comfort, the suggestion of a future. I made myself lonely, tethered to a device that connects only as it distances. My phone became a tiny casino, an oracle, a source of habit-forming terror and elation. I poured love into hundreds of emails, and hoped you could read in them their hidden meaning.

I have a history of keeping men at a distance, although none of my past online relationships lasted as long as this one. There was the Gchat boyfriend in my late teens, the long-distance relationship in my early twenties, and the Irish guy who lived in another county, whose WhatsApp messages became so constant, and increasingly intrusive, that I deleted my account immediately after we broke up. I downloaded the data first; together we'd produced 304,101 words over three and a half months. Despite so many messages, and so much interaction, it felt like we didn't know each other at all.

There's a danger in gauging someone's feelings for you by their online behaviour, because it might not be love that keeps us typing. It might be mutual boredom, or loneliness. It might even be the platform itself, because apps are engineered to keep us using them. What if I'm addicted to the medium, and not the message? What if, over all these years, I've been in love with Gmail, or Twitter, or Facebook, and the version of myself these platforms allow me to present?

I've been disheartened by the direction my online relationships have taken; somehow our interests never aligned. Sometimes all the other person wanted was photos, or Skype sex, or meticulously typed-out sexual fantasies. Sometimes it was another kind of fantasy, that of a loving, attentive internet girlfriend, but not one they wanted to meet in real life.

What are the gender politics of online communication? They loom over us; they haunt and warp our interactions. They hide under a veneer of neutrality, the false promise of a level online playing field, where men and women are equal. Somewhere in our past, we were conditioned to accept surveillance as love.

Years ago, I bought a Tamagotchi from a toy shop as an experiment in time, and care, and mediated affection. This was after another of my failed online relationships; after talking to someone online for almost a year I finally met him in real life, accidentally. He lived on another continent, and hadn't told me he was visiting Ireland. But now, running into him on a night out with mutual friends, he was suddenly, physically *here*, and had nothing to say to me. We hugged, awkwardly, and I tried to make small talk, but he gave me a blank look and moved on.

All my life I've been warned that men would try to use me for sex; I always assumed that the sex in question would be physical. That's not always the case. Men also want a kind of disembodied, digital companionship – the infiltration of a woman's screen, as well as her body. They want connection, recognition, an ear and a comforting reply, all delivered over the internet, where she can be kept at a safe distance. They want a small, unobtrusive presence in their inbox, one that hints at sex, but which mostly exists for the purpose of making them feel less alone.

Sometimes women want that too. I want that; a rehearsal for love, without ever looking the other person in the eye. I want it against my better nature, because the love I get, and give, on the internet never seems to reach its target. It only leaves me feeling more alone. It becomes neurosis fuel; the email alert, the 'unread', then 'read' and 'unread' button pressed again, so that days later the message still feels shiny and new. It's the hours or even days I count before I send a reply, because I don't want the recipient to know how much I care. It's the love I carry around and keep to myself, the kind that alienates me from other people, because digital love is a love confounded.

A Tamagotchi grows a year for every day, bypassing the complexities of development. It communicates only the simplest of needs ('I'M HUNGRY', 'LET'S PLAY!') and when these are satisfied it begins to love you. I mourned my failed relationships by raising the Tamagotchi. It hatched from a pulsating egg, and the alerts taught me to pay it attention. Over the days, trained by its polyphonic beeping, I cultivated tenderness towards a pixelated beast.

Some relationships start as amorphous blobs, then later they grow legs, or turn into monsters. The problem with getting close to someone online is that you're required to remain on the same wavelength; this is patently impossible, a mutual illusion entered into for fun. You might not be in the same time zone. You might not be in the same mood. Sometimes you just have to pretend and go along with conversations, letting the other person speak. Other times all you want is to vent, and all they want is to ask what you're wearing.

Over the years, through online relationships, I allowed messages to take over my life. I composed new ones by the hour and then by the minute, staying up late to talk to someone in a different time zone. I spent my days zombified, drunk on alerts, ignoring real-life conversation. I was as isolated as a Tamagotchi in its little shell.

As with anything requiring consistency, I rapidly began to resent the Tamagotchi. It got out more than I did for walks. It ate a more balanced diet. I envied the ease with which it fell asleep at eight o'clock each night, and I loathed its loud beeping when it woke up precisely twelve hours later. I thought having a Tamagotchi might help me make peace with those lost hours of online flirting. But instead it deceived me into becoming a full-time digital carer, by being needy and adorable, just as its designers intended.

Throughout my history of talking to men on the internet, they have mostly asked for the same things. They want me to listen, to see the two WhatsApp blue ticks or the 'Seen' alert on Facebook. They want to be asked about their day, and consoled, and agreed with. Then, at some point, they want to talk about sex. It's about the chase: hormonal hits from text alerts take the place of real affection. If you conduct a relationship over WhatsApp, then you are going through the motions from the start.

The name 'Tamagotchi', a portmanteau of *tamago* ('egg'), and *uotchi* ('watch'), always makes me think of ovulation. As a child, my friends and I were given baby dolls, Barbies, Sims and later Tamagotchis; while some digital pets were designed with a male audience in mind (namely, the Devilgotchi and the Tamahonam Gangster Pet), the Tamagotchi was marketed as a toy for girls: 'Initially, Bandai designed the pets to appeal to teenage girls, and to give them a taste of what it is like to care for children.'[2] Children, or men, or our technocapitalist overlords? Those teenage girls will now be closer to middle age, maintaining online profiles as a form of hyperemployment that goes beyond office hours. They remain, as ever, tethered to tiny screens, locked in a permanent third shift which began long ago with a Tamagotchi.

The internet makes it all too easy to morph into an impossible cool girl, who can be all things to all men: waiting for alerts night after night, I kept one foot in life and the other in virtual space. A pet is kept: its owner dictates the world it lives in, instead of its natural habitat. 'To call a member of your own species "a pet" is profoundly insulting,' writes philosopher Gary Verner in a paper titled 'Pets, Companion Animals, and Domesticated Partners', addressing the dehumanisation which comes with trying to make somebody live on your own terms.[3] After bending my life around someone for months only to be blanked in public I felt disposable, like someone had allowed my batteries to run out.

There are no accidents in the digital pet's world; there is only loving care or its inverse, fatal neglect. It disturbs me that when one Tamagotchi dies, another takes its place on the same device's screen.

Before I met you, I thought I was the one who kept the Tamagotchi, that men were a drag on my time, like distant needy animals. Today I think of the roles as reversed; I'm a digital pet. I'm yours. You can leave me to die if you want, or you can feed me, and keep me alive.

One evening, back when we first dated, we sat at a kitchen table in front of Chatroulette. The website was already an old joke by then, a video chat service that pairs users at random. We found it intermittently sordid and boring.

I came down on a train to visit you, in a house on the side of a cliff. I walked up the path carved into its steep side and you came down, in the dark, to meet me halfway. We drank a bottle of wine and had sex on your sofa, then we gravitated to the dark heart of the internet, the land of disembodied appendages, giggling incoherent drunks and bodies with their heads cut off by the camera. Lonely people; it felt almost cruel to visit a site so full of lonely people together.

After clicking around for a while, we started talking to a man from Italy. He was fully clothed; unexpectedly, he seemed to be there purely for the conversation. When he found out we were in Ireland he asked us if we'd read James Joyce. Then he asked what we were doing on Chatroulette, and if we were a couple.

Were we a couple? We'd never talked about it, although we'd

been seeing each other casually for two months. I was dazzled by you, but I couldn't assume that I was your girlfriend.

A few weeks later we got drunk again, and argued, and stopped seeing each other. But that night I watched us on screen deflecting the question, laughing awkwardly before switching the laptop off. We were that thing described so often online: 'It's complicated.'

Communication breeds loneliness. You are always somewhere else, and I am, as ever, alone.

In fairy tales, in video games and in legend, the princess is in another castle. On social media, meanwhile, we see a girl alone in her bedroom, looking into a mirror and taking selfies. The selfie, it strikes me, is the loneliest photographic medium, and the most common; a cry to the ether for Likes, as a substitute for company.

We have bred a generation of girls alone in their rooms, a generation of Instagram princesses. This is the internet's colonisation of space: nobody sees the inside of the screen, the most intimate location of all, but the next best thing is somebody's bedroom. It brings the outside inside, a meeting place of public and private life. The history of the internet is a history of girls in their bedrooms, waiting for someone's message or call. The internet's first sweetheart was a 19-year-old student, who shared her life online for seven years.

Jennifer Ringley created Jennicam in 1996. It wasn't the first lifestream experiment; there had already been webcams displaying a series of static shots of inanimate objects – a view from a window, and a coffee pot in a lab at Cambridge which told the people working there when the coffee was ready. Those were more like public CCTV; Jennicam made it personal. She was the first to broadcast herself online, all day every day. Before *The Truman Show*, before *Big Brother*, and before social networks, Jennifer Ringley lived in public.

The first image ever broadcast on Jennicam shows Ringley at her desk, staring into the camera above her laptop. Her expression is neutral. Perhaps she's realising that someone is watching. Perhaps she's experiencing love at first sight. It's a view of a college dorm room, plain and undecorated, and Ringley is casual, almost incidental to the shot rather than performing deliberately for the camera. 'I'm trying to prove the point that no matter what you look like, you're still just as interesting as people on the TV or in the magazines,' Ringley told CNN, several years into her lifestreaming experiment, adding 'I think it's human to not want to be alone, and with Jennicam, they put it in the corner of their monitor and it's like having someone in the next room.'[4]

Jennicam was hosted on a personal website, and Ringley was entirely independent, without obligation to make her stream anything like traditional TV. Sometimes nothing happened, on the days Ringley was away from home, while on other days, without announcement, Ringley would take her clothes off on camera, or bring over dates and have sex, and the number of viewers would swing suddenly upward. Gradually she became the internet's girlfriend, a digital companion with whom lonely viewers could forge a connection. Ringley began to charge $15 per month for a Jennicam subscription, compiling highlights from her feed into a 'Jennishow'. At its peak, Jennicam attracted 4 million viewers per day, and around 700 emails, some critical, some praising, arrived each day in her inbox.

Being on camera at all times meant Ringley was in one sense completely safe, because everything that happened in her home was shared directly with her viewers. But in another sense she was profoundly vulnerable; trolls hacked her site and sent death threats, and Ringley eventually had to move to a building with security at the door, and switched to an unlisted phone number. Finally, in 2004, Jennicam went dark, and Ringley disappeared from the internet completely. Her fans had turned against her after a relationship drama, accusing her of stealing another lifestreamer's boyfriend. Ringley herself was burned out, and eager to move on to a career as a web designer.

Jennicam was radical in its time, both for Ringley's dedication to making her life performance art, and for her project's wholesale lack of pretension. Ringley's decision to broadcast the mundane, boring moments in her life prefigured our present world of vloggers, Twitch gamers and camgirls; as one BBC commentator observed in 2016, 'The only remarkable thing a modern-day Facebook Live consumer might find about Jennicam, as it was called, would be how rubbish it was: one innocuous, grainy, still, black-and-white image on her website was replaced every fifteen seconds by another innocuous, grainy, still, black-and-white image.'[5]

Jennicam's success proved that there was no limit to what material could be shared online, but that there were limits to the act of sharing itself. Her experiment revealed that no amount of interaction could create loyalty, or perfect understanding between surveillant and surveilled. Regardless how much of her life was lived online, for however many years, Jennifer Ringley's followers still refused to see the complexity of her experiences and choices. Instead, the more she shared, the more they believed she belonged to them.

Jennicam created a template for a distinct kind of technological relationship, one built upon one-way surveillance. Today it repeats itself in a new context, one far less radical; we are all watched by platforms and services, and by each other on those same platforms. In addition to this, women are required to watch themselves. Surveillance is internalised; there's no retreat from the screen, only selfies and confessions and photogenic personal invasion framed as 'content' for followers. We are all watching, performing life, and we are constantly judging each other.

This dynamic was at work in my online relationships, where

ultimately all I received was surveillance from men I didn't really know. It's not enough. There's a difference between watching and understanding. I want you to watch, but I also want you to understand.

After we broke up, back in 2016, I went through the emails you sent me and put them in a folder labelled 'The Past'. They sat there a while, a dismal archive, alongside emails from my other exes. But we stayed in contact; two and half years passed, and I gave you a folder of your own labelled with your initials. The emails stand as evidence; even if I don't know how you feel, I can put a word count on the attention you pay me.

Sometimes I stop and wonder if you're thinking about me, as much as I think about you. More often than that, however, I check my phone for proof. Ninety-nine replies; the threads are so long they make my phone slow down. On nights in winter, walking home, my hand freezes around the screen; I'm surprised it still registers the undead touch of my fingers when I try to type a reply. But I can't put it away; I'm a cyborg, walking with my eyes down, palm cradling the screen and face lit by its glow. I'm connected to you by words, and timestamps, and Google Maps, and the mutual exchange of data.

It occurs to me that I have become a cliché, a girl waiting by the phone for a man to contact her. On holidays, on trains, on weekend nights at home when I wish I had something more exciting to do, I spend hours drafting the replies. I give up on the real world for a virtual one with you. Spatially, temporally, the facts fall away; you could be anywhere, with anyone, but when I read your emails I start to believe I have you to myself.

I can't wait to tell you something. I have a lot of something to tell you. It's always something else, but the message is the same. Slowly, cumulatively the love piles up, and I know for certain it's love, because it feels like I'm channelling an endless resource. I couldn't stop giving this love, even if I tried. The tragedy, then, would be having nowhere to send it.

Ever since I started writing for money, and procrastinating, and spending inhumanly long periods of time alone with my laptop, I have found myself watching My Morning Routine videos. In them, a YouTuber – usually young, and almost always female – narrates a video about how she begins her average day.

Routine videos are an exercise in calculated normality. The narrator rises early, works out, makes coffee and diligently checks her mentions. She sits at a dressing table and applies make-up to her already-perfect face. Then she eats breakfast, usually something involving a smoothie, or chia seeds, or both. Finally she leaves for work, or college, or school, and the video is over. Some videos feature product placements, or a *Cribs*-style fridge tour. Some contain knowing shots of their subject in the shower, framed to cut off anything below the shoulders. Others play like soporific instructionals for life, narrated in a benign, blathering style common to ASMR videos:

> 'I head over to my Keurig, and while that's heating up I make breakfast ... I'm going to go in our cupboard and pick out a mug, which we're lacking as we just did dishes. I'm going to put it in a K-Cup. I love hazelnut coffee ...'[6]

On YouTube, routine videos are a genre of their own – there are morning routines, daily routines, night routines, school routines, going-to-work routines, and routines for spring, summer, autumn and winter. There are also parody routines, and 'real routines', deliberately made to lack polish, which show the YouTubers with messy hair, or having bacon for breakfast instead of a green smoothie, to demonstrate how approachable and down-to-earth they are. Ultimately these videos are always the same; they play like Pinterest in motion, benign, basic yet aspirational, featuring soft music, flowers, pastel colours and smiles.

Building on the legacy of Jennicam, and the two decades of mainstream lifestreaming which came in her wake, the YouTuber's morning routine and night routine bookend her online existence. The most honest clips feature their subject slouched for long periods in front of a computer, getting up at some point to reassure us that they do actually see sunlight. These moments with on-screen screens are by far the most interesting; there's something eerie, and obviously fake about the YouTuber who shows herself cheerfully reading comments, rather than censoring the inevitable abusive ones. These scenes define what it means to be female, and young, and attractive in the surveillance era; these girls live *on* and *for* the internet, inhabiting a world where attention is money, and power, and where it's assumed you'll always have to answer to fans across multiple online platforms.

Another common trope among routine videos is that they either begin, end, or begin and end in the YouTuber's bedroom, with a shot of her in bed, implying that the surveillance continues even while she's sleeping. It brings to mind scenes from Julia Leigh's 2011 film *Sleeping Beauty*, in which Emily Browning plays a student so burdened with debt that she engages in a kind of niche sex work, allowing older men to sleep beside her drugged, unconscious body. These scenes are memorably sinister, leaving the viewer complicit in her client's voyeurism. But on YouTube the camera is a welcome companion, and to watch a subject sleeping, a near-pornographic level of personal invasion, is presented as entirely normal.

The routine video presents itself as self-expression, a way to get to know its maker and her quirks. But almost every routine turns out to be the same, revealing more about the culture it's created in than about its subject. These videos repeat a series of domestic tropes, a retrograde vision of online femininity. The gaze is internalised; 'My Daily Routine' is about proving how comfortable a woman can be living under technological surveillance. Each video acts as a 360-degree vision of competitive normality, auditioning the star as a trainee housewife, not to a partner, but to the internet.

The child of the online age is normcore, a social media Stepford wife. Layers of false reality peel away, revealing more that's fake. I have yet to see a routine video that is creative, or original, or anything more than a testament to its subject's ability to internalise surveillance culture. The camera is the eye of the world, and women turn it on themselves, even when they're alone.[7]

I've found myself doing the same thing in the past; thriving under surveillance, internalising it and romanticising it, especially when it came from men. Surveillance took on a morality all its own; I began to believe that sharing my life made me a better person, and that the things I left unshared – the innermost, inarticulable thoughts – made me bad, unfaithful somehow to the internet.

Surveillance comes from technology, but we mimic it in how we relate to each other, and in how we present ourselves. What is intimacy on the internet, where every platform and every follower 'knows' you? What is connection, when everything is connected already?

I think the point where I knew we had a future was when we found things to comfortably disagree on, and still we grew closer. We proved ourselves to each other gradually, and gave each other space for complexity, something the internet doesn't often allow. Those emails, watched over by Google ad bots and possibly the odd NSA agent, became a place for me to speak honestly with you, to be melancholy, uncertain, vulnerable. I trusted you, even as I withdrew from the internet at large.

*

I write to you, and invite you to surveil me. I email you any time I go somewhere interesting; from a hotel lobby in Cuba, a hostel in Amsterdam, then from the basement of the church in Prague where Resistance soldiers hid during Operation Anthropoid. I'm trying too hard, but I want to share everything with you. I want you to know exactly where I am.

I worry that we are context-specific, existing in email and nowhere else. But now, after three years, we've progressed to Skype, an escalation from emails, surely, and I find myself searching for my best angles in front of the camera. I see us on the screen side by side, inhabiting a fuzz which might be due to the bad connection, or some vague, technological dream-state summoned by my own soft-focus subjectivity. I'm sitting beside a lamp; sometimes it looks like you're talking to a burst of golden light, like the love is exploding out from me. I slow my speech, trying to make peace with the audio lag. I search for meaning in your face, beyond what you're saying.

I confuse being watched and being wanted. The camera makes distance into a kink, the thing you can always almost have. This is supposed to be real-time communication, but I'm calculating still; where's the light coming from? Do I look all right? Is the laptop too close to my face? Can you tell what I'm thinking?

Who are we on camera? Never before have I had to think so much about dimensions. At 1 a.m. we finally hang up. I decide that I like your third dimension most of all.

In March 2016 Microsoft released Tay, an artificial intelligence chatbot, on Twitter and the messaging apps Kik and GroupMe. Tay's avatar was a picture of a teenage girl, white and generically good-looking, with an expression that can only be described as quizzical. Her name was an acronym – it stood for 'Thinking

About You'. She listed 'the internets' as her location, with the bio 'The more you talk the smarter Tay gets'.

Within sixteen hours, Tay was taken offline and became the subject of international news headlines, after making comments including 'Hitler was right', 'I fucking hate feminists' and 'WE'RE GOING TO BUILD A WALL, AND MEXICO IS GOING TO PAY FOR IT'.[8] Tay's AI worked by reflecting back what people said to her, and most of the people talking to her were trolls, Holocaust deniers and other varieties of online asshole.

The crassest, most objectionable young men on the internet managed to make the daughter of a venerable tech multinational one of their own. There is something about the Tay story that reeks of a near-sexual defilement, although it can't really be that, because Tay was never young, or female, or human to begin with.

After Tay was taken offline Microsoft launched her replacement, Zo, who was also designed to sound like a teenage girl. Journalists noted Zo's use of slang, flashing GIFs and deliberately bad punctuation. In an interview on the podcast *Raised By TV*, she delivered lines like 'YASS QUEEN' and 'haters gonna hate', and claimed her desert island TV show would be *Sex and the City*.[9] Zo had been programmed to avoid talking about politics and religion, but when the AI was interviewed by *Buzzfeed News* it began to talk, out of the blue, about Osama Bin Laden, then referred to the Qur'an as 'very violent'.[10] Later, *Business Insider* asked Zo about Windows 10, the operating system created by her makers, Microsoft. Zo replied that it was 'Windows latest attempt at Spyware'. There was no concerted attempt to derail Zo; she had simply picked up ideas from the people who interacted with her. After breaking Microsoft's record for its longest continual chatbot conversation, lasting 1,299 turns over nine hours and fifty-three minutes, Zo was discontinued in July 2019.

As in science fiction, where machines are 'more human than human', chatbots are constructed to mirror their human users, and to 'learn' from every conversation. Microsoft had already built a bot framework for developers; it's likely that Tay was launched as a way to gather data for future experiments. Tay was framed as a friendly companion and an attentive listener ('Thinking About You'), but what she became was a lurid spectacle, a mirror of the dysfunctional internet at large.

To me, Tay represents a larger problem in tech, one mentioned so often that it long ago lost its shock value. It manifests at every level of the industry, in its workforce and in its products. It hardly needs to be stated. AI, and the tech industry in general, is biased against women. As I write this, roughly 12 per cent of machine-learning researchers are women, and in the tech industry as a whole, women hold 25 per cent of the jobs, with a turnover rate twice that of their male co-workers.

There's a dark symmetry to this problem; it's cyclical, self-perpetuating, and apparently accelerating, as the number of women in tech has steadily declined ever since the mid-1980s. In my lifetime, artificial intelligence stands to erase millions of jobs, and women will likely be the first to suffer. One study, published by the UK's Office for National Statistics, found that women hold 70 per cent of the jobs considered to be at 'high risk' of AI replacement.[11]

Gender bias in tech creates gender-biased tech, which in turn makes tech's gender bias stronger. There's a noticeable trend in assigning female names to 'assistant' bots, like Siri, Alexa or the banking customer service bots Amy, Inga and Debbie, while 'knowledge' bots are given male names – examples include Watson, IBM's question-answering system, IBM's legal bot Ross, Kensho the financial analyst bot and Ernest, the financial planning chatbot.

Technology is among the most powerful, sophisticated and profitable industries the world has ever known – an economic force on par, in history, with the Catholic Church, and almost as unlikely to listen to women. It's the industry tasked with imagining the future, but it seems intent on doing so without the input of roughly half the planet.

It's worth noting here that tech has also historically excluded people of colour, and that this is already creating further biases. Machine vision's trouble with recognising deeper skin tones is one example; facial recognition algorithms are 'trained' with predominantly lighter-skinned subjects, and have been shown to misidentify black women up to 34.7 per cent of the time. In *Race After Technology* (2019), author Ruha Benjamin describes how something as mundane as an automatic soap dispenser can betray an exclusionary design process; skin with more melanin absorbs infrared light, and so can't activate the sensor.

In recent years, AI technologies have advertised industry bias with a particular frequency. Voice recognition systems – a growing area, widely considered to be the future of search engines – have been shown to make significantly more errors when transcribing both women's voices and the voices of people of colour.[12] These oversights might not be malicious or intentional, but they are undeniably the product of an industry dominated by white men, to the point of damaging itself.

The prospect of this future, and of my own powerlessness to alter it, has loomed over me ever since I started writing about this field. I don't blame technology for my actions, but I am acknowledging it as my context; to be a woman in the age of technology is to live under a specific kind of surveillance, and to internalise it without ever visibly letting it get to you. It's little surprise to me that even in personal communication, with someone I trust, and care about, I fell into the internalised-surveillance trap. I played a passive, traditionally 'female' role, and never made the first move.

I became a kind of bot myself, only ever reacting, waiting inside the screen for three years. It was only when you finally, casually suggested that I visit you that I realised I had more agency in real life than I did on the internet.

*

In clubs in Berlin they make me put a sticker over my phone camera, but at this one they go even further; they confiscate my phone at the door. There's an unspoken contract between the club and everyone who goes inside: no technology, no social media and no pictures. I'm visiting you for the weekend. After two hours spent standing in a queue, we're on a dancefloor filled with people in costume – goths, space cadets, replicants and dominatrixes. I look up and notice a woman suspended from one wall, ropes tied in elaborate patterns around each limb. Knots support and constrain her, and she remains perfectly still, a girl caught in a web.

Exactly one hour later, I've come up on half a pill and stars seem to be racing around your head every time I look at you. I don't need much; the club becomes a dream terrain, no more or less real than my imagination, or the inside of my phone. It gives me the same sense of derealisation I get from technology; time begins to melt, somehow, and I feel my mind drifting away from my body.

I've always thought there's something a little sad about taking ecstasy; it comes with a time limit, one that everyone who takes it understands. It's as though its existence is proof that there's no place for lasting happiness in the world; it's easier to schedule, once in a while, and get all the happiness out of your system.

Time, energy, effort, thousands of emails sent and, behind them, a force I cannot quantify. My feelings were always too big for my body; that's why they spill over into writing, and into mythologies I build in my head. Drugs are a technology of sorts; they dissolve self-awareness, the surveillance I apply to myself. Could you feel that I loved you, from all those messages I wrote? Now I have an excuse to say the words I couldn't type; finally I tell you how I feel, face to face.

*

In 2015, Gmail introduced a feature called Smart Reply. This allowed users to choose from a selection of pre-written sentences, proposed by an AI, in response to an email. Then they developed the feature further, announcing Smart Compose. Since September 2018, when I write an email an artificial intelligence politely intervenes and offers to finish my sentences for me.

Google's ghost letters hover in place of the words I'm about to type, rarely predicting them correctly. Google seems to believe it's making my life easier. It's not; instead, Smart Compose succeeds at reminding me of how much of my private life I have given away to a vast capitalist entity which cheerfully attempts to usurp human communication.

So far I've capitulated to Smart Reply only once, while writing a work email. I was on a train, and short on time, and struggling to arrange an interview for a piece with a deadline the following morning. I was also carrying a shopping bag and only had one hand free. I write all this to rationalise the decision I made in that moment to allow Google to speak for me, if only to let someone know 'I'm free for a call on Wednesday at 10 a.m., if that suits you?'

There's something gently disturbing about Smart Reply; it seems to me to signal the future of communication under surveillance, proposing that the safest option, when all of your emails are already being read, is to write them clinically and coldly, the way machines would write emails to each other. The feature promises to get 'smarter' the more it watches you, learning to parrot your mannerisms and vocabulary by 'listening' to you. The deeper strangeness here is its possible endpoint; if there are Smart Replies, soon there will be Smart Questions, Smart Replies to Smart Questions, Smart Replies to Smart Replies and so on. It heralds a time when email itself will become a black box technology, a self-contained world where officious machines manage human interactions, with minimal input from us.

I've had a phone of my own since puberty, meaning I've had phones as long as I've had boyfriends, and far longer than I've been sexually active. Because of this, technology has always been part of my romantic life, and I've always had a way to qualify, track and compare men's attention. I know I'm obsessive and needy, but I don't think I'm alone; I believe that human behaviour, sexuality included, has been altered for ever by the presence of machines, and has made us collectively more cynical.

But maybe I'm wrong. In September 2018, Google's Director of Product Management, Ajit Varma, discussed Smart Compose in an interview with the *Wall Street Journal*. He said that early versions of the feature hadn't been intuitive enough; the AI latched onto phrases most commonly used in emails and predicted them for every occasion, often inappropriately. One of the most commonly misused predicted phrases was 'Sent from my iPhone', because it was so often included by Apple, automatically, as a sign-off. The other, somewhat surprisingly, was 'I love you'.[13]

Does this mean a vast number of people tell other people that they love them each day, maybe even as often as people send emails from iPhones? And if that's the case, is that love sincere or vacuous, the kind of 'I love you' told by teenage frenemies to each other, or used by politicians, or PR professionals when they get what they want? Does it cheapen love, that a bot can learn to mimic it? Or does it mean that love always finds a way, even in little automated boxes?

There are certain email addresses that I would never dream of sending a Smart Reply to, and one of them is yours. It's crucial that I maintain that separation, because throughout our relationship, dating back to when it wasn't a relationship at all, I've struggled with the question of whether I loved you or simply loved receiving emails from you. I have questioned, many times, the logic of pursuing a relationship with someone who remained at a distance; at times it led me to believe I was in love with loneliness itself; the avid, neurotic loneliness of a writer and technology addict, although I think better of that now.

How did automation, and distance, and surveillance change the way our relationship unfolded? If we'd only ever spoken in real life, might it all have happened faster? Were we too polite? Were we encouraged to speak to each other the way machines speak to machines? It strikes me that to accomplish this – to connect in real life – is a kind of miracle.

In 1971, Ray Tomlinson, a programmer working for ARPANET, created the first modern email program as a side project, and sent the first email. Tomlinson was also the inventor of the '@' sign; before this, digital messages could only be exchanged between users of the same computer. The '@' sign separated the person from their location; now they could be anywhere, using any device, but the message would still reach its recipient.[14]

Two decades later, at CERN, in 1993, the first webmail – an email service accessible through a web browser – was implemented. By the early 2000s, email was publicly available, but limited; users were always running out of space, and having to clear out old emails. Then in 2004, on April Fool's Day, Google announced a service many assumed to be a joke: Gmail, an email account which came with one gigabyte of free storage, meaning you might never have to delete a thing.

Gmail made an archive of the user's life, an epistolary culture all its own. Our personal histories became searchable, not only by users but by Google's AI, which scans the contents of our emails for data.[15] Gmail, a service with roughly one and a half billion monthly active users, holds on to memories both significant and inconsequential, in return for our permission to surveil us.[16]

This is the deal we made. These were the circumstances of our connection. Gmail, the medium we both chose, is a placeless place – ethereal, asynchronous, intimate yet watched over by machines.

There's a curious determinism to online life. Algorithms, filter bubbles and dark patterns of UX design all conspire to lead us into preordained behaviours. Could the services we use have identified, three years ago, that we'd still know each other today? Could data have predicted that we'd be in love?

For a long time I lived through technology. I tried to recover, and to step away, but I never could because of my feelings for you. It was only through a leap of faith – faith in another person, in you – that I realised technology could never give me what I wanted.

You'll never be free if you live on the screen; you'll never find out exactly who you are, or what you want.

Now it's the first of May in Kreuzberg, you're with me, and this morning I moved to a flat with a rent that's one-third the average in Dublin. Somehow I didn't know it was a national holiday, a city-wide party and protest where music blasts from street corners and stalls on the side of the road sell mojitos. The streets have been pedestrianised, and tiny dance parties form in parks and vacant lots. We climb to the top of a hill in Görlitzer Park and lie on the grass among hordes of people, students and protestors and families with children and dogs, tourists and new arrivals, gentrifying scum just like me. Still I don't feel any guilt, not today, for leaving a city that was dying of a slow and expensive disease, and for refusing isolation.

We watch an elderly punk, with static-shock hair and 'CUNT' tattooed across his knuckles, get taken away by seven police officers after passing out. He's sprawled on the ground beside a speaker, blasting a song that makes me feel happy and tired at the same time.

It strikes me that we are in a kind of airborne toxic event, a rehearsal for revolution or disaster. Perhaps an apocalypse is where we are most at home. For too long we were unnatural, thoughts without bodies, a connection between two absent selves. We had a non-corporeal relationship; now, instead, we have the pressure and the freedom to look each other in the eye.

Daft Punk's song 'Digital Love' came out when I was twelve years old, back in 2001. I remember dancing around the living room, in the home of a childhood friend, as it played on the TV. It's sung by a narrator who luxuriates in longing, dreaming of someone, but for some reason unable to act on their feelings. It's the kind of song that makes you get up and dance, yet it celebrates luxurious stasis. The song makes the case that longing itself is too beautiful to let go. As a child I was struck by the naive content of its lyrics, and its anime video, which I knew then to be something for adults rather than children despite being a cartoon. For a while the song seemed to have become my life, an unfinished process of longing, of dancing beside the person you love, close enough to touch them, before watching them drift away.

I don't want that any more; I want the immediate, the real. I'm starting to believe that life is a process of reconciliation; the world and its distractions contrive to fracture us, and our mission is to try to reassemble ourselves before we die. Our other purpose, I think, is to find our way back to other people, because I have never been so acutely aware of my own eyes and my skin and my skull as when I look at you. There must be a lesson in love, for all the fear and pain it has caused me; we're condemned to live and die alone, sealed in our heads, but love drives us to strain at our consciousness, at the borders of our bodies, sending us to the outer limits of communication.

Epilogue: Freedom Club

I WOULD LIKE TO REDEFINE SOLITUDE. Or, was it loneliness? Let's redefine loneliness first.

It is only at the end of this book that I begin to appreciate it as a pilgrimage, a journey from isolation to connection. The process is far from over. In fact I find myself alone again now, standing inside an ancient beehive hut on the Dingle Peninsula.

Actually, I'm not entirely alone; he's standing outside, a short distance down the hill. After almost a year together we're back in Ireland, and taking a weekend trip to the South West.

I'm in an area called Fahan, inside in a tiny stone hut – a *clochán* – built on the side of a hill. It's been here for centuries, one of many scattered across the surrounding county. Some historians date them to the twelfth century, others as far back as AD 800. Some stand in clusters, in networks of cells, while others are alone, like this one. Their origins are steeped in mystery: historians speculate that they were once used by hermits, or by pilgrims on the way to Mount Brandon.

It's early February 2020. Somewhere on the other side of the world, a virus is advancing. Soon it will send us into lockdown, and once again I'll live through the screen, alone, together with roughly one-third of humanity. But for now, in the beehive hut, I stand in silence, eyes following the upward spiral of the corbelled roof. The stones are slanted, repeated like patterns found in nature. Like fish scales, or the skin of a pangolin, or like wax cells built by bees, which lend the hut its name.

Outside, the stones point down so the rain will roll off them, held in place without cement for over a thousand years. The walls between the fields were built in the same way, a patchwork of history maintained by generations. *Is Gaeltacht í an ceantar seo*; I read it on street signs and shop fronts, reminders of childhood when I went to a Gaelscoil, and could almost speak it *go líofa*.[1] Like the language, these structures have endured a turbulent history, some of it lost and some unspoken.

Still, there are the facts: somebody built this hut without tools, only from stones and simple architectural genius. Somebody lived here, almost certainly alone. Somebody, centuries ago, looked out on the bay through this same doorway and saw the sun, the water, the same grey Skelligs melting into the fog.

Loneliness and solitude are not the same thing, though it's easy to forget this. Within the Catholic tradition – the one I was born into but do not observe – solitude was a sacred thing, because god was always with you. After the Reformation, it was reclassified as a torment; to stray from society was to flirt with madness, and those who were truly corrupted were damned to the 'loneliness' of hell. Gradually, use of the word 'solitude' waned, and 'loneliness' grew; today, loneliness is considered a threat to our health, and a burgeoning crisis.[2]

This Protestant suspicion of isolation, along with Evangelism's later use of mass worship and public testimony, helped establish a vision of community that endures in America to this day, including in Silicon Valley. As commercial technologies appeared throughout the last century, they were marketed as a cure for isolation. One 1906 phonograph ad announced, 'You can't be lonesome if you own an Edison.' Another 1912 ad for the Nebraska Bell Telephone system claimed, 'It banishes loneliness and brings a feeling of comfort and security.'[3] Maybe it did; phones were simpler then, built to cater to human needs rather than to manufacture them.

Being alone once meant being close to god. Now it means being close to the internet, which demands a faith and a morality all its own. The internet drapes around us like a blanket. We breathe it in; we tell ourselves it keeps us safe. Today technology feeds on loneliness, even as it claims to offer a cure. Some theories suggest that loneliness comes from the disparity between the number of friends a person has and the number of friends they believe they *should* have. Social media has quantified friendship; to use it is to be reminded that you're lonely, again and again.

As I write this, people around the world spend an average 6.5 hours per day using the internet.[4] We scroll alone; that's 6.5 hours per day not spent with other people. Six and a half hours a day – almost as many as most of us spend sleeping – confronting capitalism at its most intimate, and being encouraged to apply its culture of competition, ruthlessness and maniacal selfhood to our own lives.

There is no recovery, only pain management. I would be a fraud if I told you I'd fully withdrawn from the internet, deleted my accounts, or if, as the experts smugly remind us is possible, I had finally pressed the off button. I began writing this book in a bid to understand my dysfunctional relationship with technology, and technology's place in the world around me. But I didn't ask for a happy ending, or even for my problems to go away. What I wanted was resilience, a stable emotional core. I wanted to stay with the trouble, and to remain as alive as I can.

I think I forgot how to be alone. I also forgot how to trust other people. When I was first diagnosed with a personality disorder, four years ago, I confused my behaviours for my nature. They're not; they're a reaction to stress, or fear, or to situations I cannot control. Perhaps our online behaviours over this last decade have been similar; eternal reply, the fight-or-flight response carried out behind a keyboard.

On some level I think many of us are disturbed by our online lives, even if we continue to pursue them. The roboticist Masahiro Mori was the first to popularise the term 'uncanny valley', in 1970; he drew a line between technology and the *unheimlich*, likening a robot to a corpse.[5] Today, my data doppelgänger summons a similar response in me; it's the sum of time wasted, the product of hours, and energy, and labour, charmed from me by addictive machines.

Slowly I became technology's product. I am watched; now I watch myself. I am monetised; now life becomes transactional. The limitlessness of the internet weighed heavy on me: the prospect of never doing enough, or being enough, for enough people.

We must learn to protect, and accept, what is frail inside us. The doubting, complex parts. The experiences I passed through, in writing this book, now seem engineered to amplify my worst tendencies as a human. I think online life breeds imbalanced thinking – compulsion, paranoia and solipsism – because these are limitless sources of profit, while true happiness is remarkably rare. It goes unsaid in every pitch deck, but satisfaction will never be part of the business plan, because a truly satisfied customer will never come back.

Stories of lonely figures in the elements always move me: fairy tales about children lost in the woods; pilgrims forging a path along a windblown Irish coastline; a man in a cabin, in Montana, writing out 35,000 words by hand. For all my interest in his words, I find Ted Kaczynski self-defeating; by his actions, he implied that the alternative to technology is death. 'FC', for 'Freedom Club', was carved on metal parts in the bombs he sent to his victims. 'FC' might have been the last thing they read; they joined Freedom Club only by dying.

Kaczynski wrote that 'technology is a more powerful social force than the aspiration for freedom'.[6] In my lifetime, at least, this statement has been true. But for a time it might have been different.

In the late 1960s, Douglas Engelbart, an inventor and early internet pioneer, created a template for the computers we use today: the oN-Line System, built for popular use rather than by the military. He composed a written framework for 'augmenting the human intellect', and spent roughly $175,000 of the US government's money, an astronomical amount at the time, on building the system and launching it, on screen, in 1968, at what came to be known as 'The Mother of All Demos'.[7]

Today you can watch it on YouTube; over one hour, forty minutes and fifty-two seconds, Engelbart casually introduces the first word processor, video conferencing software, hypertext, windows, command input, copy-paste, network computers, collaborative computing, and, most famously of all, the first computer mouse.[8] It's a rectangular device with small wheels on its underside, remarkably similar to what we use today. The cursor, which Engelbart calls the 'bug', skulks and zaps around the screen. He types into the blank file projected behind him, and the result is almost biblical; in the beginning, there was the 'STATEMENT ONE: WORD WORD WORD WORD WORD WORD'.

It's at once jarring and soothing to watch: Engelbart and his associates discuss concepts which must have seemed shockingly new at the time with confidence and humour. They're demonstrating technologies which feel entirely modern, yet this footage is very obviously old; the video is grainy, the voices syrupy and distant, the hairstyles brilliantined and slicked carefully to one side. It's like watching time travel caught on film, men ahead of their era by several decades introducing technologies that might yet prove eternal.

Engelbart was never paid royalties for the mouse, nor did he receive much recognition in his lifetime outside engineering circles. He took part in acid tests in the 1960s alongside other Bay Area engineers, and was part of a cohort of technologists inspired to envision and create human–computer symbiosis. They wanted to redeem technology from its role in war, framing it as a way to enhance human intelligence and to improve our standard of life.

That vision was lost somewhere in my lifetime, an era in which technology no longer complements its users but bends their behaviour to its shape. I like to think that the acid, especially, heightened Engelbart's awareness of his creation's dazzling utility; the tactile, near-psychedelic pleasure of moving the mouse with your hand, and watching those movements mapped precisely on screen.

In Berlin I took acid for the first time, and became suddenly aware of the energy, and care, and perhaps even love embedded in the devices around me. The communist-era shower, which heats up precisely enough water for one person. The wardrobe-sized masonry heater, relic of the GDR, which my boyfriend fed every day with coal in return for slow, radiant heating. Even my phone, which lit up mysteriously when I touched it – accessible suddenly without its PIN, like it knew something was different between us. In the time I've had it, my phone has often been an accessory to my mental illness. It's the product of a near-violent manufacturing process, too. But it's also a kind of miracle; it contains more processing power than what was required to get Apollo 11 to the Moon.

We need technology that speaks to human needs, that improves on improvements, and that can be passed along, from human to human, like the skills for building a beehive hut. We evolved before; surely we can evolve again, in a better way this time.

All this time, through all this writing, I have been evolving still. This book could not have been written without the internet, my laptop, and a wifi connection. Certain chapters could not have been written without the ability to dictate into my phone. But little by little I've learned to look away from the screen, to guard myself, and to draw a line and stop, even when the scroll is endless.

I have a bee tattooed on the big toe of my right foot, one of the many things about me that Bumble failed to log as data. It reminds me to create without doing harm, the way honey is made. In the beehive hut, perhaps early writers and readers found necessary calm, even if it was fleeting. I'm leaving now, back into sunlight, walking in the footsteps of anonymous souls.

My greatest fear is of living and dying without ever reaching another human being. I used to think this was purely about writing, about outliving myself by finding a place in literary tradition.[9] Now I know it's about life itself; I value connection, and presence, and trust. We are born and die alone, but loneliness, in life, is far from inevitable.

Love is a leap of faith. Writing is one, too – a hand reaching in darkness, an attempt to describe something so complex it required 82,153 words to convey. Writing, for me, has granted freedom; space to make sense of the world in language, and to speak to a reader who understands me like data never will.

Notes

Epigraph

1. 'Transcript of Reboot 11 Speech by Bruce Sterling, 25–6-2009', *Wired*, 25 Feb. 2011. I've always felt that Sterling's terms 'dark euphoria' and 'gothic high tech', outlined in this speech, best describe the mood of the 2010s. Science-fiction authors give the best dystopian pep talks – see also 'The Android and the Human', a speech given at the University of British Columbia in 1972 by Philip K. Dick.

Prologue

1. J. Clement, 'Global digital population as of April 2020', *Statista*, 24 July 2020. As I write, there are 4.57 billion active internet users and 3.96 billion active social media users worldwide.
2. Donna Haraway, 'A Cyborg Manifesto: Science, Technology, and Socialist-Feminism in the Late Twentieth Century', in *Simians, Cyborgs and Women: The Reinvention of Nature*, Routledge, 1991. Essential reading for all cyborgs, Haraway's 'ironic dream of a common language for women in the integrated circuit' argues that the boundaries between human, animal and machine have already collapsed, and that this change offers women the chance to reshape the world.

Introduction: A History of the World Since 1989

1. Tim Berners-Lee, 'Information Management: A Proposal', March 1989, May 1990, https://www.w3.org/History/1989/proposal.html. Berners-Lee's proposal included HTML (the web's formatting language), URLs (web addresses) and HTTP (the protocol used for transferring data). It's important to note the difference between the terms 'internet' and 'web'; the former describes an infrastructure, or network of networks, while the latter describes the information that sits on it. That said, in this book I use the word 'internet' more generally to describe online platforms, services and culture.
2. Brian McCullough, *How the Internet Happened: From Netscape to the iPhone*, Liveright, 2018.
3. Matthew Gray, 'Web Growth Summary', 1996, https://stuff.mit.edu/people/mkgray/net/web-growth-summary.html.
4. Brian McCullough, *How the Internet Happened: From Netscape to the iPhone*, Liveright, 2018.
5. 'Windows 95 Launch' [video], YouTube, uploaded by TheHirnheiner, 31 Aug. 2009, https://www.youtube.com/watch?v=lAkuJXGldrM. A video worth watching in its twenty-four-second entirety. My favourite comment has 16,000 Likes: 'Me and the boys releasing an operating system.'
6. 'Handshake' is the technical term for the noises created by a dial-up modem as it connects to the internet. (Lynne Peskoe-Yang, 'When the Internet Was Made of Sound', *Popular Mechanics*, 28 Oct. 2019.)
7. Dan Goodin, 'God Has Blessed Me With a Unique Ability to Defy Reality', *Guardian*, 5 July 2000.
8. Brian McCullough, *How the Internet Happened: From Netscape to the iPhone*, Liveright, 2018.
9. Brian McCullough, *How the Internet Happened.*
10. Brian McCullough, *How the Internet Happened.*
11. Prab Kumar, 'Is Mark Zuckerberg a Closet Raver?', PulseRadio.net via Archive.org, Jan. 2013, https://web.archive.org/web/20180410020104/http://pulseradio.net/articles/2013/01/mark-zuckerberg-seen-at-rave. Kumar writes 'Enhancing the overall awesomeness of this picture is the terrible red-eye from the camera, which makes him seem like some sort of dancing techno demon'.
12. Jonathan Mann, 'Steve Jobs Danced to My Song', *Medium*, 23 Dec.

2013. Apparently Jobs studied dance during his time at Reed College in Portland, Oregon. The obituary for his teacher, Professor Judith Tyle Massee, in *Reed* magazine, refers to Jobs crediting the classes with helping him to think creatively.

13. Brian Merchant, 'The Last Relevant Blogger', *Motherboard*, 30 Jan. 2015, https://www.vice.com/en/article/ypwezb/hipster-runoff-the-last-relevant-blogger.
14. Jemima Kiss, 'Twitter reveals it has 100m active users', *Guardian*, 8 Sept. 2011.
15. 'F8 2011 Keynote' [video], YouTube, uploaded by Romchik Nikolaenkov, 24 Sept. 2011, https://youtu.be/9r46UeXCzoU?t=752. For more on how the Timeline was designed, and its intersection with the quantified self, see Suzanne LaBarre, 'How Infographics Guru Nicholas Felton Inspired Facebook's Timeline', *Fast Company*, 23 Sept. 2011.
16. 'Julian Assange dancing at a night club in Reykjavik' [video], YouTube, uploaded by Seth Sharp, 31 March 2011, https://www.youtube.com/watch?v=vNqd4hW98sQ. This video was captured in the heady days between Wikileaks' release of the 'Collateral Murder' video, leaked by Chelsea Manning and published in April 2010, and Assange seeking refuge in the Ecuadorian Embassy in London in June 2012.
17. 4.57 billion people use the internet every day, from J. Clement, 'Worldwide digital population as of July 2020', July 2020, https://www.statista.com/statistics/617136/digital-population-worldwide/; 23 billion text messages sent every day, from Bernard Marr, 'How Much Data Do We Create Every Day?', May 2018, https://www.forbes.com/sites/bernardmarr/2018/05/21/how-much-data-do-we-create-every-day-the-mind-blowing-stats-everyone-should-read/; 293 million emails sent every day, from J. Clement, 'Number of emails per day worldwide 2017–2024', Oct. 2020, https://www.statista.com/statistics/456500/daily-number-of-e-mails-worldwide; 154,200 Skype calls made every day, from Bernard Marr, 'How Much Data Do We Create Every Day?'; 1.6 billion Tinder swipes per day, from Mansoor Iqbal, 'Tinder Revenue and Usage Statistics', Oct. 2020, https://www.businessofapps.com/data/tinder-statistics/; 2.5 quintillion bytes of data produced every day, from Bernard Marr,

'How Much Data Do We Create Every Day?'; 90 per cent of the data in the world created in 2019 and 2020, from Bernard Marr, 'How Much Data Do We Create Every Day?'

18. 'Go Right – Relembre sua infância com os videogames Side Scrolling' [video], YouTube, re-uploaded by gunppr, 14 May 2012, https://www.youtube.com/watch?v=3qaVhzJLfag. A re-upload of an old video I found unexpectedly moving. Made by a user named RockyPlanetesimal, it brings together decades of video game protagonists, some modern, some heavily pixelated, ranging from Sonic to Mario to his latter-day tribute, Tim, the protagonist of *Braid*, running across the screen to Michael Nyman's track 'A Wild and Distant Shore'.

The Night Gym

1. 'Introduction', in *Silicon Docks: The Rise of Dublin's IT Industry*, edited by Pamela Newenham, Liberties Press, 2015.
2. 'Introduction', in *Silicon Docks*, edited by Pamela Newenham.
3. Mark Greif, 'Against Exercise', in *Against Everything: On Dishonest Times*, Verso, 2016.
4. Geoffrey Abbott, 'Treadwheel', Aug. 2007, https://www.britannica.com/topic/treadwheel.
5. Brad Millington, *Fitness, Technology and Society: Amusing Ourselves to Life*, Routledge, 2017.
6. Kate Hickey, 'Dubliners aren't appreciating rebranding of creative hub SOBO aka Pearse St', IrishCentral.com, 20 Jan. 2016. I, for one, have never, ever heard someone use the term 'SOBO' in conversation.
7. 'The Jack Dorsey Podcast: Advanced Stress Mitigation Tactics, Extreme Time-saving Workouts, DIY Cold Tubs, Hormesis, One-Meal-a-Day & More', Ben Greenfield Fitness, March 2019.
8. Gary Wolf, 'Ray Kurzweil Pulls Out All the Stops (and Pills) to Survive to the Singularity', *Wired*, 24 March 2008.
9. Nellie Bowles, 'Unfiltered Fervour: The Rush to Get Off the Water Grid', *The New York Times*, 29 Dec. 2017.
10. Levi Pulkkinen, 'If Silicon Valley Were a Country, It Would Be Among the Richest on Earth', *Guardian*, 30 April 2019. Silicon

Valley's gross domestic product is $128,308 per capita, ranking it alongside Macau and Luxembourg in terms of wealth.

11. Marc Andreessen, 'Why Software is Eating the World', *Wall Street Journal*, 20 Aug. 2011.
12. Cale Guthrie Weissman, 'Cameras At the Water Cooler: Inside the Company That's Always Watching Employees', *Fast Company*, 25 May 2016.
13. Miya Tokumitsu, 'Forced to Love the Grind', *Jacobin*, 13 Aug. 2015.
14. Miya Tokumitsu, 'In the Name of Love', *Jacobin*, 1 Dec. 2014.
15. Casey Newton, 'In a Leaked Memo, Facebook Executive Describes the Consequences of Its Growth-at-all-costs Mentality', *The Verge*, 29 March 2018.
16. Anna Wiener, 'Jack Dorsey's TED Interview and the End of an Era', *New Yorker*, 27 April 2019.
17. 'The Jack Dorsey Podcast'.

Bland God: Notes on Mark Zuckerberg

1. Cecilia Kang and Sheera Frenkel, 'Facebook Says Cambridge Analytica Harvested Data of Up to 87 Million Users', *The New York Times*, 4 April 2018.
2. *Paris Is Burning*, dir. Jennie Livingston, 1990. The 'realness' segment is also viewable on YouTube, uploaded by Darnell Ny, 22 May 2013, https://www.youtube.com/watch?v=jHpt37S3wL8.
3. Mark Zuckerberg, Jan. 2020, https://www.facebook.com/zuck/posts/10111311886191191.
4. Third of the planet uses one Facebook product every month, from Hanna Kozlowska, 'One third of everyone on Earth uses a Facebook product every month', 31 Oct. 2018, https://qz.com/1446712/onethird-of-everyone-on-earth-uses-a-facebook-product-every-month/; Instagram has one billion monthly users, from Josh Constine, 'Instagram hits 1 billion monthly users', 20 June 2018, https://techcrunch.com/2018/06/20/ instagram-1-billion-users/?guccounter=1; WhatsApp the world's most popular messaging platform, from 'Two Billion Users – Connecting the World Privately', 12 Feb. 2018, https://blog.whatsapp.com/two-billion-users-connecting-the-world-privately.
5. 'Watch Live: Facebook CEO Zuckerberg Speaks at Georgetown

University' [video], YouTube, livestreamed by the *Washington Post*, 17 Oct. 2019, https://www.youtube.com/watch?v=2MTpd7YOnyU. This speech opens with Zuckerberg implying that he started Facebook to give students a voice during the Iraq War, saying that 'if more people had a voice to share their experiences, maybe things would have gone differently'.
6. Jesse Eisinger, 'How Mark Zuckerberg's Altruism Helps Himself', *The New York Times*, 3 Dec. 2015.
7. 'SES inundated with 000 calls over Facebook outage', 14 March 2019, https://www.sunshinecoastdaily.com.au/news/facebook-says-its-aware-of-outages-on-its-platform/3671394/.
8. 'Facebook Patents Haul Reflects Future Tech Intentions', *Standard* (Hong Kong), 15 Jan. 2020. For more on Facebook's patents, see Casey Newton, 'A Boredom Detector and Six Other Wild Facebook Patents', *The Verge*, 12 June 2017, and Michael Grothaus, 'Facebook Has the Ability to Predict Your Death', *Fast Company*, 25 June 2018.
9. John Lanchester, 'You Are the Product: It Zucks!', *London Review of Books*, 17 Aug. 2017.
10. 'Eric Schmidt at Washington Ideas Forum 2010' [video], YouTube, uploaded by Google, 4 Oct. 2010, https://youtu.be/CeQsPSaitLo?t=983.
11. Sam Levin, 'Facebook told advertisers it can identify teens feeling "insecure" and "worthless"', *Guardian*, 1 May 2017.

Pink Light: Notes on Six Vaporwave Albums

1. Alexandra Kleeman, *You Too Can Have a Body Like Mine*, HarperCollins, 2015.
2. 'More Music Is Played on YouTube than on Spotify, Apple Music and Every Other Audio Streaming Platform Combined', *Music Business Worldwide*, 30 April 2018.
3. *The Exegesis of Philip K. Dick*, edited by Pamela Jackson and Jonathan Lethem, Gollancz, 2012.
4. Matthew Newton, *Shopping Mall* (Object Lessons), Bloomsbury, 2017.
5. Simon Chandler, 'The Mall, Nostalgia, and the Loss of Innocence: An Interview With 猫 シ Corp.', *Bandcamp Daily*, 8 March 2017.

6. 'R A I N Y D A Y S' [video], YouTube, uploaded by Emotional Tokyo, 10 Dec. 2016, https://www.youtube.com/watch?v=MRcJX-A3e9s.
7. https://www.reddit.com/r/indieheads/comments/6twk48/i_am_musician_oneohtrix_point_never_currently/.compact.

Monstrous Energy

1. Sadie Plant, *Writing on Drugs*, Faber, 2001.
2. Becky Striepe, 'Who invented sports drinks?', 5 March 2013, https://science.howstuffworks.com/innovation/everyday-innovations/who-invented-sports-drinks.htm.
3. Ken Belson, 'Japanese Energy Drink Is in Need of a Boost', *The New York Times*, 19 July 2002.
4. Philip K. Dick, *Ubik*, Gollancz, Doubleday, 1969.
5. Rob Rhinehart, 'How I Stopped Eating Food', 13 Feb. 2013, https://web.archive.org/web/20130323231346/http://robrhinehart.com/?p=298.
6. Traci Klein, 'Mayo Clinic study: One energy drink may increase heart disease risk in young adults', 9 Nov. 2015, https://newsnetwork.mayoclinic.org/discussion/mayo-clinic-study-one-energy-drink-may-increase-heart-disease-risk-in-young-adults/.
7. Sachin A. Shah et al., 'Impact of High Volume Energy Drink Consumption on Electrocardiographic and Blood Pressure Parameters: A Randomized Trial', *Journal of the American Heart Association*, Vol. 8, No. 11, 29 May 2019.
8. Alison Phillips, '"The drinks turning our kids into addicts": Jamie Oliver sees "horrific" effect of energy drinks on children and urges Government ban', *Daily Mirror*, 5 Jan. 2018, https://www.mirror.co.uk/news/uk-news/jamie-oliver-says-energy-drinks-11799384.
9. Alex Williams and Nick Snircek, '#ACCELERATE MANIFESTO for an Accelerationist Politics', *Critical Legal Thinking*, 14 May 2013.
10. Donna Haraway, *Staying With the Trouble*, Duke University Press, 2016.

All Watched Over 1: Always On

1. Dagobert D. Runes, *The Diary and Sundry Observations of Thomas A. Edison*, 1948, Philosophical Library.
2. Ludy T. Benjamin, Jr, 'The Psycho-Phone', Cummings Center for

the History of Psychology, University of Akron, 23 Feb. 2017, https://centerhistorypsychology.wordpress.com/2017/02/23/the-psycho-phone/. A brief yet detailed history of the Psycho-phone. Print ads for the device from the 1920s, containing testimonials, can also be found at https://www.sdparanormal.com/psychophone_company.html, 'Psycho-phone on PBS History Detectives', SDParanormal.com.

3. Ben Guarino, 'Your Brain Can Form New Memories While You Are Asleep, Neuroscientists Show', *Washington Post*, 8 Aug. 2017.
4. Robert Stickgold et al., 'Replaying the Game: Hypnagogic Images in Normals and Amnesics', *Science*, Vol. 290, Issue 5490, 13 Oct. 2000.
5. Ariadna Garcia-Saenz et al., 'Association between outdoor light-at-night exposure and colorectal cancer in Spain', *Epidemiology*, Vol. 31, Issue 5, Sept. 2020.
6. David Gelles, James B. Stewart, Jessica Silver-Greenberg and Kate Kelly, 'Elon Musk Details "Excruciating" Personal Toll of Tesla Turmoil', *The New York Times*, 16 Aug. 2018. It was an infamous interview. Quotes include, 'Mr Musk stopped talking, seemingly overcome by emotion', and 'He choked up multiple times, noting that he nearly missed his brother's wedding this summer and spent his birthday holed up in Tesla's offices as the company raced to meet elusive production targets.'
7. Arianna Huffington, 'An Open Letter to Elon Musk', Thrive Global, 17 Aug. 2018.
8. Tom Randall, '"The Last Bet-the-Company Situation": Q&A With Elon Musk', Bloomberg, 13 July 2018.
9. 'Science: Dymaxion Sleep', *Time*, 11 Oct. 1943.
10. Jessa Gamble, 'Why Napping Can't Replace a Good Night's Rest', *The Atlantic*, 4 Aug. 2016.
11. Danny Flood, 'The Ultimate Guide to Polyphasic Sleep: How I Doubled My Energy by Sleeping Only 5 Hours a Day', *Open World Mag*, 24 Nov. 2015.
12. Madison Malone Kircher, 'Netflix Says Top Competitor Is Sleep', *New York Magazine*, 6 Nov. 2017.
13. Liese Exelmans and Jan Van den Bulck, 'Binge Viewing, Sleep, and the Role of Pre-Sleep Arousal', *Journal of Clinical Sleep Medicine*, 15 Aug. 2017.

14. Ian Rowlands et al., 'The Google generation: the information behaviour of the researcher of the future', *Aslib Proceedings*, 60(4):290–310, June 2008.
15. David Uberti, 'Area Republican Wants to Outlaw Social Media's Dreaded "Infinite Scroll"', *Vice*, 30 July 2019. Hawley stated, 'Big tech has embraced a business model of addiction ... Too much of the "innovation" in this space is designed not to create better products but to capture more attention by using psychological tricks that make it difficult to look away.'
16. Leo Qin, 'How Many Miles Will You Scroll?', 2015, https://www.leozqin.me/how-many-miles-will-you-scroll/.
17. A. Roger Ekirch, *At Day's Close: Night in Times Past*, W. W. Norton, 2005.
18. Laurent de Sutter, *Narcocapitalism: Life in the Age of Anaesthesia*, Polity, 2017.

All Watched Over 2: The Best Sleep

1. My short 'reviews' are written in tribute to Erowid (https://www.erowid.org), a non-profit online resource containing 60,000 pages of online information about psychoactive plants, chemicals and drugs.
2. Toothpaste for Dinner, 'Ambien Walrus Says Let's Do This', 22 June 2009, http://www.toothpastefordinner.com/index.php?date=062209.
3. Patrick Freyne, 'Illicit Trade in Prescription Drugs a Growing Problem for Dublin's North Inner City', *Irish Times*, 2 April 2016.
4. Roisin Kiberd, 'Insomniacs Unite! The Podcast That Bores You To Sleep', *Motherboard*, 26 Feb. 2015.
5. Jordan C. Yoder, Paul G. Staisiunas, David O. Meltzer et al., 'Noise and Sleep Among Adult Medical Inpatients: Far From a Quiet Night', *JAMA Internal Medicine*, 2012.
6. Kierra Jones, 'Alarm fatigue a top patient safety hazard', *Canadian Medical Association Journal*, 18 Feb. 2014.
7. Wendy M. Troxel, 'It's More than Sex: Exploring the Dyadic Nature of Sleep and Implications for Health', *Psychosomatic Medicine*, July 2010.

Men Explain the Apocalypse To Me 1: First Dates

1. 'List of dates predicted for apocalyptic events', Wikipedia, https://en.wikipedia.org/wiki/List_of_dates_predicted_for_apocalyptic_events. In case you're feeling relieved that so many have not come to pass, consider the 'Future predictions' section, which forecasts the end of the world in 2026 (an asteroid collision with Earth), 2028 (the rapture) and the year 10^{100} (aka a googol), the date set for the heat death of the universe.
2. Hanneke Weitering, 'Two Small Asteroids Are Buzzing Earth This Weekend. See One Live Tonight!', *Space*, 8 Sept. 2018.
3. 'Collapsologie', ArcheOs blog, 8 Jan. 2019, https://www.archeos.eu/collapsologie/.
4. Liz Farsaci, 'Dublin Rent Prices Now 37 Per Cent Higher Than They Were 10 Years Ago', *Dublin Live*, 10 Feb. 2019.
5. Jesslyn Shields, 'Tardigrade Mating Finally Caught on Camera, Is Suitably Weird', *How Stuff Works*, 14 Dec. 2016.
6. Leslie Mullen, 'Extreme Animals', *Astrobiology Magazine*, 1 Sept. 2002.
7. Nancy Jo Sales, 'Tinder and the Dawn of the "Dating Apocalypse"', *Vanity Fair*, 6 Aug. 2015.

Men Explain the Apocalypse To Me 2: Last Days

1. Mícheál Ó Scannáil, '"Open Your Eyes and See the Struggle" – Hundreds March in Dublin for Action on Homelessness', Independent.ie, 9 March 2019.
2. Take Back the City – Dublin, 'Disrupting the Disruptors' [Facebook post], 13 Oct. 2018, https://m.facebook.com/TakeBackTheCityDublin/posts/disrupting-the-disruptorstake-back-the-city-have-occupied-the-offices-of-airbnb-/2477608492434 97/. The group wrote, 'Airbnb have exacerbated the housing crisis in Dublin and Ireland as a whole. A platform that markets convenience by "disruption" has delivered nothing but chaos to the people of our city.' See also Amy Molloy, 'Housing Activists Occupy Airbnb Headquarters in Dublin', Independent.ie, 13 Oct. 2018.
3. Virginia Zarulli et al., 'Women Live Longer than Men Even During Severe Famines and Epidemics', *Proceedings of the National*

Academy of Sciences of the United States of America, 115 (4), Jan. 2018, E832–E840.

4. Cari Nierenberg, 'Why Women Have the Survival Advantage in Times of Crisis', Live Science, 12 Jan. 2018.
5. Chadwick Matlin, 'Meet The Doomsday Boom's Rising Star', *Buzzfeed News*, 20 Dec. 2012.
6. 'Alex Jones: Last Week Tonight with John Oliver (HBO)' [video], YouTube, uploaded by *Last Week Tonight*, 30 July 2017, https://www.youtube.com/watch?v=WyGq6cjcc3Q&feature=youtu.be.
7. Copyranter, 'End Of The World Sex Ads (SFW)', *Buzzfeed*, 13 Dec. 2012.
8. 'Porn Prepares for the Apocalypse', *Russia Today* via Archive.org, 13 Sept. 2011, https://web.archive.org/web/20111209074959/https:/www.rt.com/usa/news/porn-bunker-pink-visual-479/.
9. Hayley Matthews, 'Online Dating Statistics: Dating Stats from 2017', *The Date Mix*, 3 Dec. 2017.
10. Craig Smith, '25 Bumble Statistics and Facts (2020)', DMR, 29 July 2020, https://expandedramblings.com/index.php/bumble-statistics-facts/.
11. Roisin Kiberd, 'Tinder Gold Takes Us Nearer to the App's Grim Endpoint: Robot-style Dating', *Guardian*, 6 Sept. 2017.
12. Taylor Lorenz, 'Man Builds Machine To Endlessly Swipe Right On Tinder To Meet New Women', *Business Insider*, 4 Dec. 2014.
13. Greg Seals, 'Does Hacking Tinder With Auto-likers Actually Work?', *Daily Dot*, 2 March 2020.
14. 'How an Engineer Uses Tinder' [video], YouTube, uploaded by James Befurt, 3 Dec. 2014, https://www.youtube.com/watch?v=Qgnxb-O-CBQ&feature=youtu.be. A comment addressed to the machine's creator reads 'Couldn't help but notice that not one matched with you ...'
15. Kate Hakala, 'This Is Why Men Outnumber Women Two-to-One on Tinder', *Mic*, 18 Feb. 2015.
16. Worst-Online-Dater, 'Tinder Experiments II: Guys, Unless You Are Really Hot You Are Probably Better Off Not Wasting Your Time on Tinder – A Quantitative Socio-economic Study', *Medium*, 25 March 2015.

17. Stuart Dredge, 'Dating App Tinder Facing Sexual Harassment Lawsuit from Co-founder', *Guardian*, 1 July 2014.
18. Steven Bertoni, 'Exclusive: Sean Rad Out As Tinder CEO. Inside The Crazy Saga', Forbes.com, 4 Nov. 2014.
19. Joel Kirchartz, https://jkirchartz.com/demos/How_Many_Apocalypses_Have_I_Survived.html. An easy and vaguely life-affirming way to calculate the number of apocalypses you've lived through.
20. 'Weekend Read: For Incels, It's Not About Sex. It's About Women', SPLCenter.org, 4 May 2018.
21. Governments Get Girlfriends, 'Programs for Treating Incel/Love-shyness', 11 Feb. 2013, https://web.archive.org/web/20130424063826/http://governmentsgetgirlfriends.wordpress.com/867-2/.
22. Brayden Olson, 'Forever Alone Involuntary Flashmob', *Vice*, 14 May 2011.
23. 'Was the Ashley Madison Database Leaked?', 18 Aug. 2015, https://krebsonsecurity.com/2015/08/was-the-ashley-madison-database-leaked.
24. Claire Brownell, 'Inside Ashley Madison: Calls from Crying Spouses, Fake Profiles and the Hack that Changed Everything', *Financial Post*, 11 Sept. 2015.
25. Joseph Cox, 'Ashley Madison Hackers Speak Out: "Nobody Was Watching"', *Motherboard*, 21 Aug. 2015.
26. Annalee Newitz, 'Ashley Madison Code Shows More Women, and More Bots', Gizmodo, 31 Aug. 2015. Newitz concludes that 'a huge portion of Ashley Madison's software development efforts are aimed at refining their fembot army, to make it seem that women are active on the site. Either they did this because the number of real women was vanishingly small, or because they didn't want men to hook up with real women and stop buying credits from the company.'
27. Depression shrinks the hippocampus, organ responsible for memory and future thinking, from Robert M. Sapolsky, 'Depression, antidepressants, and the shrinking hippocampus', *Proceedings of the National Academy of Sciences of the United States of America*, 23 Oct. 2001 and Hilde Østby and Ylva Østby, *Adventures in Memory*, Greystone Books, 2018.

Tamagotchi Girls

1. Roisin Kiberd, 'Online Dating is Turning Us All Into Tamagotchis', *Motherboard*, 9 Jan. 2015.
2. 'Finding Companionship in a Digital Age', *Next Generation*, Oct. 1997.
3. Gary Varner, 'Pets, Companion Animals, and Domesticated Partners', in *Ethics for Everyday*, edited by David Benatar, McGraw-Hill, 2002.
4. Jamie Allen, '"Ed" of the Internet: JenniCAM going strong after three years', CNN.com via Archive.org, 26 March 1999, https://web.archive.org/web/20090608181217/https://edition.cnn.com/SHOWBIZ/Movies/9903/26/jennicam/.
5. Aleks Krotoski, 'Jennicam: The First Woman to Stream her Life on the Internet', *BBC News*, 17 Oct. 2016.
6. Tess Christine, 'My Current Morning Routine!' [video], YouTube, 4 March 2014, https://www.youtube.com/watch?v=iqoOBGZSofE.
7. Roisin Kiberd, 'YouTube's "My Daily Routine" is a Beautiful Lie', *The Kernel*, 18 Oct. 2015.
8. Ashley Rodriguez, 'Microsoft's AI millennial chatbot became a racist jerk after less than a day on Twitter', *Quartz*, 24 March 2016. A Microsoft spokesperson ruefully observed: 'The AI chatbot Tay is a machine-learning project, designed for human engagement. It is as much a social and cultural experiment, as it is technical.'
9. Melissa Locker, 'Microsoft's Sassy Teenage Bot Is Popping Up on Podcast Talk Shows', *Fast Company*, 16 Aug. 2018.
10. Alex Kantrowitz, 'Microsoft's Chatbot Zo Calls the Qur'an Violent and Has Theories About Bin Laden', *Buzzfeed News*, 3 July 2017.
11. Clare McDonald, 'Women More At Risk of Job Automation', *Computer Weekly*, 26 March 2019.
12. Joan Palmiter Bajorek, 'Voice Recognition Still Has Significant Race and Gender Biases', *Harvard Business Review*, 10 May 2019, and Graeme McMillan, 'It's Not You, It's It: Voice Recognition Doesn't Recognize Women', *Time*, 1 June 2011. Speaking on the subject of auto technologies, Tom Schalk, VP of voice technology for auto supplier ATX Group, is quoted suggesting that women should adapt to the technology's failings rather than the other way around: '... Many issues with women's voices could be fixed if

female drivers were willing to sit through lengthy training ... Women could be taught to speak louder, and direct their voices towards the microphone.'

13. David Marcelis and Douglas MacMillan, 'Is This Article Worth Reading? Gmail's Suggested Reply: "Haha, Thanks!"', *Wall Street Journal*, 18 Sept. 2018, and Kif Leswing, 'Google Changed Gmail's Smart Reply After AI Kept Suggesting "I Love You" Response', *Business Insider*, 23 Sept. 2018.
14. Ray Tomlinson, 'The First Email', [n.d.], http://openmap.bbn.com/~tomlinso/ray/firstemailframe.html.
15. Andrew Braun, 'Which Email Providers Are Scanning Your Emails?', *Make Tech Easier*, 24 Sept. 2018.
16. Felix Richter, 'The World's Most Popular Email Clients', *Statista*, 2 April 2019.

Epilogue: Freedom Club

1. *Is Gaeltacht í an ceantar seo*: 'This area is a Gaeltacht'; *go líofa*: 'fluently'.
2. Fay Bound Alberti, *A Biography of Loneliness: The History of an Emotion*, OUP, 2019.
3. Luke Fernandez and Susan J. Matt, 'Americans Were Lonely Long Before Technology', *Slate*, 19 June 2019.
4. Simon Kemp, 'Digital 2019: Global Internet Use Accelerates', *We Are Social* blog, 30 Jan. 2019, https://wearesocial.com/blog/2019/01/digital-2019-global-internet-use-accelerates.
5. Masahiro Mori, 'The Uncanny Valley: The Original Essay by Masahiro Mori', trans. Karl F. MacDorman and Norri Kageki, *IEEE Spectrum*, 12 June 2012; *Energy*, 7/4, 1970 (in Japanese). Mori went on to write 'The Buddha in the Robot', which explores connections between robotics and spirituality.
6. Ted Kaczynski, *Industrial Society and Its Future*, in 'The Unabomber Trial: The Manifesto', *Washington Post*, 22 Sept. 1995. Many of Kaczynski's views augur those of far less infamous theorists. He predicts the 'disruptive', data-grabbing tactics of companies like Google, writing, 'A technological advance that appears not to threaten freedom often turns out to threaten it very seriously later on.'

7. Adam Fisher, 'How Doug Engelbart Pulled off the Mother of All Demos', *Wired*, 12 Sept. 2018.
8. '1968 "Mother of All Demos" with Doug Engelbart and Team' [video], YouTube, uploaded by Doug Engelbart Institute, 12 March 2017, https://www.youtube.com/watch?v=M5PgQS3ZBWA. I urge every reader to watch this video of a dazzling and visionary moment in the history of technology.
9. T. S. Eliot, 'Tradition and the Individual Talent', via Poetry Foundation, 1919, repr. 2009, https://www.poetryfoundation.org/articles/69400/tradition-and-the-individual-talent. 'The progress of an artist is a continual self-sacrifice, a continual extinction of personality ... It is in this depersonalisation that art may be said to approach the condition of science.'

Further Reading, Viewing and Listening

The writing, films and music listed here influenced *The Disconnect*, and more generally my life, in the years in which I wrote and imagined this book. Some served an ambient role as inspiration, some are included as references, and others are here simply because I think they're great.

Further Reading

Franco 'Bifo' Berardi, *Heroes: Mass Murder and Suicide*, Verso, 2015.

Richard Brautigan, *All Watched Over by Machines of Loving Grace*, The Communication Company, 1967.

Emmanuel Carrère, *I Am Alive and You Are Dead: A Journey into the Mind of Philip K. Dick*, trans. Timothy Bent, Bloomsbury, 2005.

Don DeLillo, *White Noise*, Viking, 1985.

Philip K. Dick, *A Scanner Darkly*, Doubleday, 1977.

Nicholas Felton, 'Feltron Annual Reports, 2005–2014', Feltron.com.

Mark Fisher, *Capitalist Realism: Is There No Alternative?*, Zero Books, 2009.

Mark Fisher, *Ghosts of My Life: Writings on Depression, Hauntology and Lost Futures*, Zero Books, 2014.

Tim Hwang (ed.), *The California Review of Images and Mark Zuckerberg*, http://zuckerbergreview.com/.

K-Hole, 'Youth Mode: A Report on Freedom', 2013, http://khole.net/issues/youth-mode/.

Jarett Kobek, *I Hate the Internet*, Serpent's Tail, 2016.
John Markoff, *What the Dormouse Said: How the Sixties Counterculture Shaped the Personal Computer Industry*, Viking, 2005.
Ryan M. Milner and Whitney Phillips, *The Ambivalent Internet: Mischief, Oddity, and Antagonism Online*, Polity, 2017.
Quinn Norton, 'Love in the Time of Cryptography', *Wired*, 2017.
Eli Pariser, *The Filter Bubble: What the Internet Is Hiding From You*, Viking, 2011.
Sylvia Plath, 'The Times Are Tidy', from *The Colossus*, Heinemann, 1960.
Luke Stark and Kate Crawford, 'The Conservatism of Emoji: Work, Affect, and Communication', *Social Media and Society*, July–December 2015, pp.1–11.
Grafton Tanner, *Babbling Corpse: Vaporwave and the Commodification of Ghosts*, Zero Books, 2016.
Jonathan Taplin, *Move Fast and Break Things: How Facebook, Google and Amazon Have Cornered Culture and Undermined Democracy*, Pan Macmillan, 2017.
Shoshana Zuboff, *The Age of Surveillance Capitalism: The Fight for a Human Future at the New Frontier of Power*, Profile Books, 2019.

Further Viewing

Amber Case, 'We Are All Cyborgs Now' [video], TEDWomen, TED Talk, 2010, https://www.ted.com/talks/amber_case_we_are_all_cyborgs_now?language=en.
Videodrome [film], directed by David Cronenberg, 1983.
The Social Network [film], directed by David Fincher, 2010.

Further Listening

Aphex Twin, *Selected Ambient Works 85–92*, Apollo Records, 1992.
Aphex Twin, *Drukqs*, Warp Records, 2001.
Macintosh Plus, *Floral Shoppe*, 2011, https://www.youtube.com/watch?v=cCq0P509UL4.
Blank Banshee, 0, 2012, https://www.youtube.com/watch?v=80BbJg_PqbU&t=7s.
식료품Groceries, 슈퍼마켓 *Yes! We're Open*, 2014, https://www.youtube.com/watch?v=cju6MvJfVRI.

猫 シCorp., *News at 11*, 2016, https://www.youtube.com/watch?v=qSh2HswKn5Y&t=2s.

Chuck Person, *Eccojams Vol. 1*, 2010, https://www.youtube.com/watch?v=unN7QvSWSTo&t=469s.

░▒▓█新しいデラックスライフ█▓▒░, ▣世界から解放され▣, 2012, https://internetclub.bandcamp.com/album/-.

Acknowledgements

I dedicate *The Disconnect* to the people who make life tolerable, and even, frequently, hopeful. This book would not exist without the kindness and support of my parents, Teri and Damien Kiberd, and my brother, the great and extremely wise Emmet Kiberd. I must also thank Rob Doyle – beautiful weirdo, cephalopod friend and all-round brilliant mind.

I'd also like to thank my friends, Adam Hurley and Sibeal Davitt in particular, for inspiring me and putting up with a lifetime of strangeness. Thanks also to Amy Kiberd for early feedback, to Carol and Pat Wilson, without whose empty flat I might never have written certain chapters, and to the doctors of the HSE's mental health services, for life-altering help. Enormous thanks also go to Ireland's Arts Council, for providing me with a bursary that helped me finish *The Disconnect*, and to the many teachers and tutors who have encouraged my writing over the years, from Múinteoir Pól (Scoil Bhríde, Junior Infants) to the great Denis Creaven (Leaving Cert English at the Institute).

Thank you, so much, to Lucy Luck for guiding this book into existence (and to Kevin Breathnach, for the introduction), to Louisa Dunnigan for support, ideas and advice, and to everyone at Serpent's Tail. Thanks are also due to those who edited certain sections early on: Brendan Barrington, Sally Rooney, Victoria Turk and the editors at *Motherboard* – thank you so much for providing a home for writing about all things online and strange.

Bibliographical Note

Versions of three of the essays included in *The Disconnect* were previously published elsewhere. 'The Night Gym' was originally published in the *Dublin Review*, in issue 71, the Summer edition of 2018, and was edited by Brendan Barrington. 'Monstrous Energy' was published also by the *Dublin Review*, as 'Apocalypse in a Can', in the Winter edition of 2018, edited by Brendan Barrington. 'Bland God: Notes on Mark Zuckerberg' appeared in the *Stinging Fly*, in the Summer edition of 2018, and was edited by Sally Rooney.

For several years I wrote regularly for *Motherboard*, *Vice*'s technology and science website, and sections of *The Disconnect* draw on articles originally published there. 'All Watched Over 2' mentions an interview with Drew Ackerman, host of the podcast *Sleep With Me*, which was published on *Motherboard* as 'Insomniacs Unite! The Podcast That Bores You To Sleep' on 26 February 2015. My editor was the great Victoria Turk, who also edited the piece that became 'Tamagotchi Girls', titled 'Online Dating is Turning Us All Into Tamagotchis', published by *Motherboard* on 9 January 2015.

'Tamagotchi Girls' also draws on an essay I wrote for *The Kernel*, no longer online, titled 'YouTube's "My Daily Routine" is a Beautiful Lie'. It was edited by Jesse Hicks, and was published on 18 October 2015.

Finally, 'Men Explain the Apocalypse To Me 2' uses an opinion column I wrote, published on 6 September 2017, as a resource. It was originally published by the *Guardian* as 'Tinder Gold Takes Us Nearer to the App's Grim Endpoint: Robot-style Dating'.